FARMALL CUB *Encyclopedia*

THE ESSENTIAL GUIDE TO MODELS, HISTORY, IMPLEMENTS, AND REPAIR

KENNETH UPDIKE AND RACHEL GINGELL

OCTANE PRESS

Contents

Foreword
THE FIRST COMPACT TRACTOR 6
by Kenneth Updike

Introduction
THE FINE ART OF TRACTOR RESCUE 8
by Rachel Gingell

Chapter 1
KNOW YOUR CUB'S HISTORY 11
Design of a New Machine 15
Building a Factory for the Cub 20
1950s Refinement of the Cub 24
Closure of the Louisville
Plant and Foundry 26

Chapter 2
CUB MODEL GUIDE 29
Original 1947–54 Cub 29
1950 White Demonstrator Cub 34
Cub Lo-Boy 39
1955–58 Cub 47
1959–63 Cub 52
1964–75 Cub 56
Industrial Cub 58
Model 154 Cub Lo-Boy 63
Model 185 Cub Lo-Boy 69
1975–1979 Cub 72
Model 184 Cub Lo-Boy 76
Foreign-Made Cubs 84

Chapter 3
GETTING TECHNICAL WITH YOUR CUB — 89
Engine — 89
Engine Oil System — 92
Fuel — 94
Carburetor — 95
Ignition — 96
Cooling System — 98
Transmission and Final Drive — 100
Axles, Wheels, and Tires — 102
Touch Control Hydraulic Lift — 104

Chapter 4
HOW TO BUY A CUB — 107

Chapter 5
CUB IMPLEMENT GUIDE — 111
Model 1000 "One-Arm" Loader — 120
Model 154, 185, 184 & Lo-Boy Implements — 121
Some 1950 Implement Pricing — 122
Fast Hitch Implements — 122
Three-Point Hitch Cub Implements — 127
Aftermarket Cub Implements — 127

Chapter 6
CUB PAINT & DECAL GUIDE — 129

Chapter 7
REPAIR AND MAINTENANCE — 135
7.1 How to Change Your Oil — 136
7.2 Lube and Grease — 139
7.3 Check the Air Pressure in Your Tires — 141
7.4 Adjust Your Brakes — 142
7.5 Adjust the Clutch Pedal — 143
7.6 Replace Your Front Seals — 144
7.7 Rebuild Your Carburetor — 147
7.8 Service Your Battery and Cables — 154
7.9 Tune Up Your Distributor or Magneto — 155
7.10 Service Your Charging System — 161
7.11 Replace Your Spark Plugs — 164
7.12 Replace the Starter Switch and Button — 165
7.13 Test Your Coil — 166
7.14 Test Your Engine Compression — 167
7.15 Adjust Your Engine Valves — 168

Appendices — 170
Serial Number Listings — 170
Acknowledgments — 172
Index — 173

FOREWORD

The First Compact Tractor

BY KENNETH UPDIKE

The Farmall Cub was the last major mechanical breakthrough to eliminate the horse and mule from the farm. This small one-plow tractor was specifically designed for small farms that were still using horses in the middle of the twentieth century. World War II caused delays in the Cub's entry to the market, which may have limited some of its sales. The tractor likely would have been a hit if available during the war, as it was ideally suited to help small farmers struggling with the labor shortages caused by the war.

The next addition to the family was the Cub Lo-Boy, an evolution of the model that opened a new market of tractor sales that International Harvester (IH) desperately wanted to penetrate: the industrial and commercial market. This adapted tractor opened new sales avenues for IH and helped spark life into Cub tractor sales.

Body style changes gave the model freshened looks that helped the Cub evolve with a changing world. The basic design of the machine didn't change radically, but the sheet metal was updated to match the other new IH tractors of the day.

I doubt IH engineers envisioned the Cub and Cub Lo-Boy to have such long production lives. Tens of thousands of these tractors can be found on the farm, factory, fairway, or garage all across America today and in many other parts of the world. The production span lasted until 1981, and the tractor is one of the most popular small tractors ever built.

The adaptation of the Cub transmission and differential assembly to form the Cub Cadet tractor only helped to bolster the reputation of these great little tractors.

New laws governing vehicle emissions passed in the 1970s, and the fact that the Cub tooling and its factory were both badly in need of an overhaul helped lead to the demise of the Cub tractor line for IH. The modern compact tractors that are so wildly popular today all owe their existence to the first compact tractor, the one that set the bar of excellence, the Farmall Cub.

We've put as much detail as possible into this book about the Cub, but so much more is out there about these popular small machines. I hope that this book helps to expand upon some of the ideas that IH had when it made these great little tractors.

INTRODUCTION

The Fine Art of Tractor Rescue

BY RACHEL GINGELL

I'm on a mission to rescue America's working tractors from dusty barns, fields, and fencerows around the nation. Welcome to my shop!

Many old tractors are still relevant sources of reliable power in the modern world. Their simple designs, iconic styles, and ease of repair make them historical treasures—much too important to toss on the scrap heap! Rescuing all of these aging tractors is far too much work for just one person, so I'm teaching mechanics (both beginners and experienced ones) how to repair their own old iron.

The Farmall Cub is a classic American tractor, renowned for its go-to capabilities around the farm. It's a great project tractor for beginners, hobbyists, and farmers looking for a dependable workhorse for small chores. Parts are plentiful, and nearly every repair can be accomplished with ordinary shop tools. While this book talks specifically about the Cub, International Harvester used many similar designs on their other Letter Series tractors. Many of the techniques I describe in this guide apply universally to Farmall tractors of the same era.

What do I know about repairing tractors? Plenty! My family has been running (and repairing) tractors for generations. My dad and grandpa started me out young, delivering tools and floor-dry around their repair shop on my tricycle. It wasn't long before I started to make simple repairs alongside them on my own tractors, buying and selling what I repaired to pay for my first truck and college tuition. Dad and Grandpa were top-notch teachers, showing me not only the mechanics, but how to have plenty of fun along the way. Their knowledge was first passed on to me, and now I'm passing it on to the world with a series of instructional repair videos. Through the power of the internet, we discovered countless tractor owners hungry for easy-to-understand information on how to repair their own machines at home.

It's my hope that this book gives you the confidence you need to buy, repair, and most of all *use* your very own piece of American farm history: the Farmall Cub.

CHAPTER 1

Know Your Cub's History

The rapid mechanization of farms and the transformation of the American landscape from a mostly agrarian nation to an industrial and urban nation drove changes in the needs for mechanized farming equipment. IH recognized that a tractor market existed for the part-time or urban farmer as well as the small rural farmer. The tractors being made in the 1920s and 1930s were too big and clumsy to fit into small fields and work in closely cultivated crops without damaging them. The farmer who owned a single horse or mule also needed a tractor.

IH devised a plan to build such a tractor for these customers. This new machine was code named Farmall X by IH's engineering department, or just the "baby Farmall" as others in the company knew it.

While most tractor companies focused only on bigger and more powerful tractors, IH didn't forget about the little guy. The Farmall X was the answer for these new potential tractor-buying customers. On December 20, 1944, IH's tractor and implement engineering divisions showed the first prototypes of the Farmall X to IH management. None of the parts used to make the Farmall X came from either the Farmall A or Farmall B, meaning some expensive tooling had to be made before tractor production could start and profits could be made.

IH had started to build smaller tractors in the 1930s aimed at reaching farmers who were still using horses. The Farmall F-12, F-14, Culti-Vision A, and AV were scaled-down versions of their "bigger brothers." These met the needs of the farmer running 50 to 85 acres, but were still at a price level that exceeded that of the typical budget for a 40-acres-or-less farmer. IH needed to get a smaller tractor yet to reach this yet-untapped market. The Farmall X (later renamed the Farmall Cub) was the tractor that reached these farmers. IH intended to market the Farmall X in the $400 range, meaning

CUB PROMOTIONAL EXHIBIT
IH used various iterations of bears and bear cubs to promote the Cub. This is the "Carousel Bear" sitting at the 1947 Harvester Farm exhibit at the Museum of Science and Industry in Chicago. *Wisconsin Historical Society*

FARMALL CUB TRACTOR SERIAL NUMBER 501
This is the very first production model Cub built, serial number 501. Note that earlier versions were built to test the idea—these are prototypes—and typically a few pre-production machines are handmade and sent out for more refined testing. The first production model is the first finished machine released to be sold to the public. IH used the number 501 as a starting serial number on most of their tractors. This historic tractor is safely restored and saved today in a private collection. *Wisconsin Historical Society*

its net manufacturing cost needed to be around $225. To achieve this cost it was initially proposed to use a two-cylinder upright engine. Ultimately, a four-cylinder engine was selected for its smoothness and torque characteristics.

The Farmall Cub tractor was essentially a two-thirds scaled version of the Farmall Super A tractor. Farmall Engineers and Marketing envisioned the Cub as the gasoline tractor replacement for the one-horse or one-mule farmer in America. The post–World War II era brought forth a new wave of potential tractor customers: folks who had a small urban acreage, or larger vegetable gardens. Tobacco growers were also targeted as a large buyer of Cub tractors.

The Cub and its descendents (Cub Lo-Boys and Cub Cadets) outstripped expectations to become popular tractors that are now icons of American farming. The Cub family had the longest production run of any IH tractor. The Cub was built from 1947 to 1979, a span of thirty-two years that saw farming and IH change dramatically. Note that the Cub did not have the largest production volume for IH; both the wildly popular Farmall H and Farmall M outsold the Cub and its descendents.

FARMALL CUB WITH MID-MOUNTED GRADER BLADE

A Cub with a mid-mounted grader blade is an effective tool to scrape and level the barnyard. The operator has a clear view of the blade and the massive trip spring that protects the tractor from blade overload. *Wisconsin Historical Society*

FARMALL CUB WITH MOUNTED CULTIVATOR

This side view of the Cub fitted with a mounted cultivator clearly shows the lifting linkage from the hydraulic lift-all unit. The cast iron front-wheel weights aided tractor steering in soft ground. Fingertip control of the cultivator was accomplished with the lift-all control lever the operator is touching. *Wisconsin Historical Society*

FARMALL CUB WITH REAR-MOUNTED BUZZ SAW

A not-so-common accessory for the Cub was a rear-mounted buzz saw (not IH-supplied). Here two IH company men—real users would not wear suits and fedoras—pose for a photo. The use of this type of saw may have been common seventy or eighty years ago, but safety regulations have since banished it to the scrap heap. *Wisconsin Historical Society*

FARMALL CUB AIRPORT TRACTOR

Who says the Cub was limited to just farm chores? Here a Cub tows a twin-engine airplane. Cub tractor sales probably increased ten-fold to airlines after this photo was taken. The Cub was never offered in a special airport tug version. The uses for the Cub tractor were nearly limitless. *Wisconsin Historical Society*

CULTI-VISION VIEW FROM THE SEAT
The unobstructed view from the driver's seat on the Cub offered no obstructions. The left pedal on the platform controls the tractor's drive clutch. The two pedals on the right control the left- and right-wheel brake. When cultivating (like here), the driver can clearly see the plants to avoid any damage to them.
Wisconsin Historical Society

Design of a New Machine

IH's engineering planning for the Cub was well underway long before World War II. During the war years, IH devoted nearly all of its resources to the output of machines, tools, and weapons, which played a huge role in winning the war. The Cub was delayed to keep the American war machine running. Whenever possible, IH engineers did sneak some "Cub time" into their schedule to keep the progress made so far alive. At the end of World War II, IH launched an all-out war of its own. The war was to get the Cub from blueprint drawings into a working tractor. IH's postwar product-expansion program budget was $150 million. More than one-third of that amount was spent on the Cub. IH hoped the Cub would extend the benefits of small-tractor farming power to a vast market never before served by tractor manufacturers. IH recognized the enormous possibilities for business represented by the buying power alone of the small-farm owners. To tap this market could bring IH millions of dollars of revenue, increased market share, owner loyalty, and more profitable dealerships.

The Farmall Cub was produced at the Louisville Works, with IH's long-established policies in mind. First, the intent was to balance the budget for design (which was considerable in this case) with the anticipated market (which was large as well for the Cub). Equally important was the goal of selling the Cub at affordable prices that would permit the widest possible use.

IH announced the Cub tractor to the public at its late 1945 summer field demonstration program held at the Harvester Experimental Farm in Hinsdale, Illinois. Even though it would take IH another two years to actually produce the tractor for public sale, IH heavily promoted the Cub well in advance of its release. This media frenzy not only caused interest to increase but also helped IH convey to the marketplace that it intended to make machines for every farm, no matter what its size may be.

When IH finally introduced the Farmall Cub to the public for sale in 1947, IH had spent a jaw-dropping $55 million for research, engineering, and machine tooling to take this tractor into production.

IH had planned to have Cub out several years earlier, but the war prevented this from happening. During the World War II years, there was a large-scale trend toward larger, commercially operated farms. IH recognized the value of keeping the family farm economically sound as one of its corporate goals. By offering the small, low-priced tractor with specially designed implements to match, IH hoped to reverse that trend.

As an aside, note that the highly anticipated trend of a growing number of smaller farms did not emerge. IH Chairman of the Board Fowler McCormick stated in 1937 that IH was to be put on the record as believing in the preservation and future of the

FARMALL CUB DEMONSTRATION
Men in jumpsuits and pith helmets on Farmall Cub tractors lined up to demonstrate the tractor and implements. The demonstration was staged in conjunction with a branch meeting in Great Falls, Montana, in 1947. *Wisconsin Historical Society*

FARMALL CUB WITH GRADER BLADE, MOVING SNOW

The grader blade for the Cub can be front mounted to allow the tractor to push snow. On this cold, sunny day, the Cub with blade is making quick work of snow drifts in the driveway. *Wisconsin Historical Society*

CUB TRACTOR WITH CULTIVATOR

The Culti-Vision feature of the Cub is ideal for cultivating corn, even check-row planted corn that can be cultivated in several directions. The check-row planting and cultivating method was common before the invention of herbicides. *Wisconsin Historical Society*

FARMALL CUB WITH PLOW

This early-model Cub tractor has no hydraulic system fitted to it. The long blue lever by the operator lifted the plow. The Cub circle decal on the radiator grille leg indicates this is a 1947 model. *Wisconsin Historical Society*

FARMALL CUB WITH PLANTER
This photo was taken at IH's Harvester Farm located in (then rural) Hinsdale, Illinois. The large white barn in the background was part of the main farm complex. County Line Road would be at the far top of the photo. The Cub tractor here with two-row planter has a steering guide rod affixed to the right-hand headlight. This was a very early form of manually operated GPS (Grip Point Steer) guidance. *Wisconsin Historical Society*

FARMALL CUB WITH VEGETABLE PLANTER
This high-clearance Cub has special tall and thin rear wheels and taller front wheels that were a special option from IH. The four-row planter fitted under the tractor is for planting small vegetable seeds.
Wisconsin Historical Society

family-sized farm. The number of small farms in America would decrease dramatically.

In fact, the hottest market in the post-World War II era was for larger machinery, and IH soon was joined by other tractor manufacturers in a race of ever-increasing tractor horsepower and size. Despite this trend, the low-priced Cub tractor was an important tool that allowed the small farmer to maintain profitability.

The small farmer had several choices for tractors in the late 1940s. The Cub's main competitor was the Allis-Chalmers Model G. The Model G was marketed more as a garden tractor than as a farm tractor. The Model G had a rear-mounted engine/transmission with mid-mounted implements. With 10 horsepower on the belt, the Model G was nearly identical in power to the Cub. Both tractors offered their own unique version of Culti-Vision, a term coined by IH. The Cub could accept either front-, mid-, or rear-mounted implements. The Model G, because of its rear-engine design, could only use mid-mounted implements. Allis built the Model G from 1948 to 1955. This was roughly the same time period as the original "mesh grille" Cub was.

Building a Factory for the Cub

IH had planned that the majority of the Cub tractors made would be sold for tobacco farming. While looking at possible factory sites to build the Cub, IH concentrated its efforts on a site near the tobacco belt in the United States. This geographical area was that encompassing the states of Kentucky, Tennessee, and the Carolinas.

With World War II still firmly gripping nearly every country around the globe in battles, IH was planning to build the Farmall Cub tractor as soon as the war ended. A new factory site located in Wood River, Illinois, was selected to be the home of Farmall Cub tractor production. Land procurement and site grading were started in 1945. In 1946, IH dropped its plans for the factory in Wood River and announced that it had purchased the former Curtis Aircraft Company factory in Louisville, Kentucky, from the War Assets Administration. This would become known as the Louisville Works, or as most IH enthusiasts call it, "Louisville." IH promised the townspeople of Wood River that eventually an IH factory would be built there, but it never materialized. The exact reasoning behind the last-minute switch to buy Louisville has not yet been discovered, but the lack of postwar materials and the fact that the Curtis

C-60 ENGINE PISTON ASSEMBLIES
Careful inspection of the C-60 engine components (like the piston assemblies here) was critical to the Cub performing well. IH built the four-cylinder, water-cooled model C-60 engine at their Louisville Works plant. *Wisconsin Historical Society*

factory may have been "too good of a deal to pass up" are the strongest ideas at this time.

After extensive retooling and remodeling of the factory for tractor production, the first IH tractor built at Louisville was shipped on April 11, 1947. Only one month later, the first Cub would finally roll down the assembly line. The Farmall Cub, the Farmall A, and the Farmall B were all produced at Louisville when it opened.

The Louisville plant was a complete manufacturing facility. It took in rough stock and produced finished product. Beginning in 1949, the plant had its own foundry that produced many hundreds of different gray iron castings, not only for its own production, but for other IH factories, too. The foundry had an iron melt capacity of almost 1,000 tons a day and was IH's largest. A modern forge shop with fourteen presses ranging in size from 1,000 tons to 8,000 tons produced engine crankshafts, gear blanks, and other forgings for many IH factories. In later years, an engine assembly line produced a dozen different basic models of four- and six-cylinder gasoline engines for IH and other companies. The six tractor assembly lines at Louisville produced not only the Cub, Cub Lo-Boy, and Cub Cadet, but also small and mid-size agricultural tractors, forklifts, and IH's line of

C-60 ENGINE ASSEMBLY
The parent bore (sleeveless), IH-built C-60 engine that powered the Cub and Cub Lo-Boy was carefully assembled using many components that IH cast or fabricated itself. Here the engine short block is being assembled. *Wisconsin Historical Society*

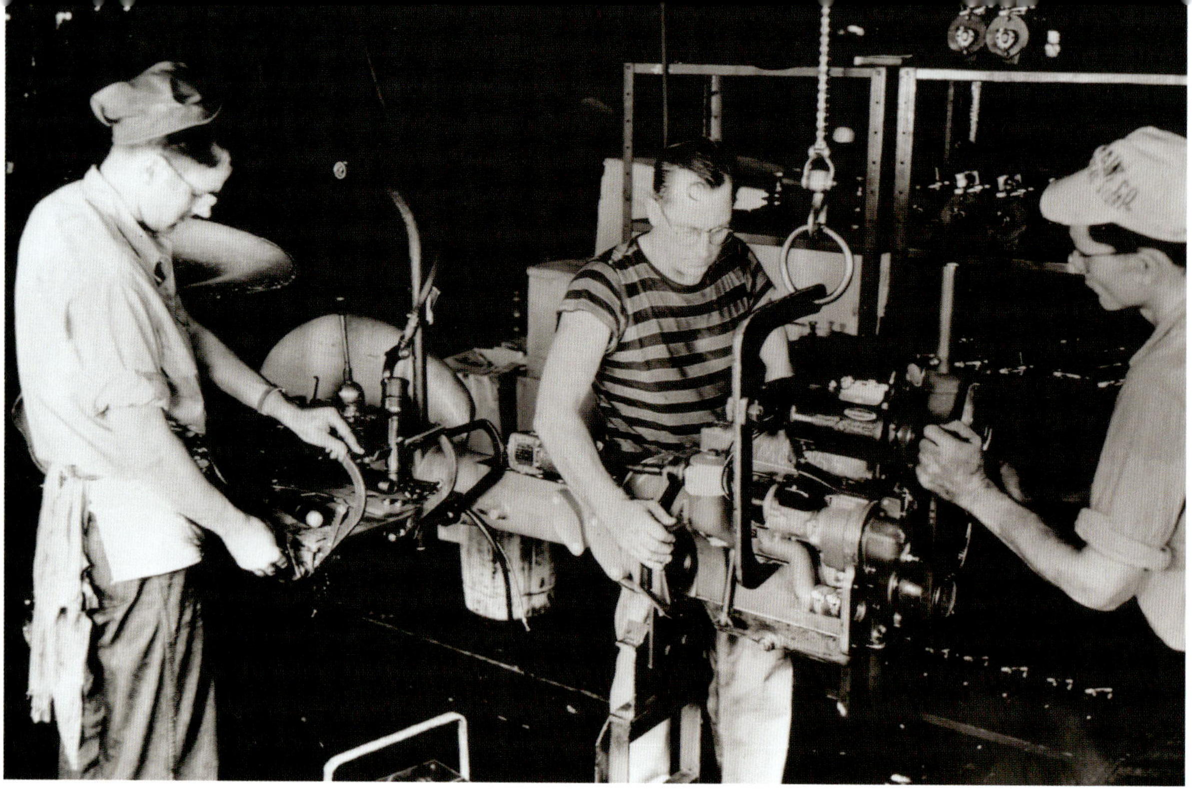

CUB TRACTOR BEING ASSEMBLED

Once the C-60 engine was joined to the drivetrain of the Cub, the assembly was soon painted and ready for delivery. Here, workers install the brake pedals on the platform and mate the engine to the drivetrain. Notice the large rack of black-painted Delco Remy generators on the shelves behind the workers.
Wisconsin Historical Society

CUB TRACTOR SUB-ASSEMBLED

The Cub tractor was hung by special fixtures from the overhead conveyor line. This nearly complete tractor has its gas tank installed and will be painted next. Notice the black-painted headlights and starter. Delco Remy supplied these to IH painted black. The headlights and starters were painted red with the rest of the tractor.
Wisconsin Historical Society

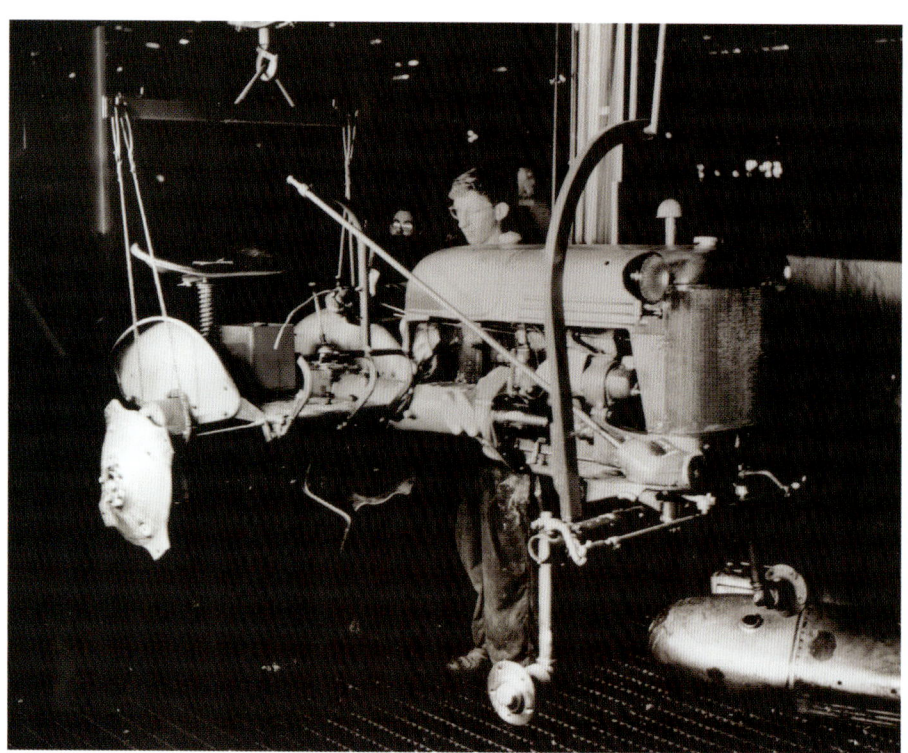

CUB TRACTOR BEING PAINTED
The Cub was nearly completely assembled when it was painted. The waterfall behind the tractor collected the oversprayed paint. The tractor was painted using a high-volume handheld paint gun. The process wasn't flawless, and factory-painted tractors were occasionally sold with a small paint run.
Wisconsin Historical Society

industrial wheeled tractors. The plant area included more than 145 acres, with 48 acres under roof.

In the three years of operating Louisville Works from 1946 to 1949, IH incurred a total operating loss of over $21.5 million. Delays in tooling the Cub sent outside parts vendors' schedules into a spiral. By the time Louisville was ready to make tractors, the parts vendors had committed to other customers. The large union labor force at Louisville (like at most other IH factories) severely limited the IH management's choices on what to do. As parts shortages became clear, only partial assembly of units could be accomplished until the missing pieces arrived.

Parts Plants

Many of IH's other factories produced component parts for the Cub. Tractor Works, West Pullman Works, and Milwaukee Works furnished forgings. The engine governor weights and raw carburetor castings were made at Tractor Works; transmission differential shafts, reverse, first-, second-, and third-speed gears and clutch shafts came from Milwaukee Works; and West Pullman Works made the rear axles, magnetos, and completed carburetors. Also, West Pullman was the source of the Cubs' screws, bolts, rivets, and roller bearings. The Rock Falls Works furnished some clips, control rods, and coiled steel springs. The grey iron foundry at the Indianapolis Works made castings for the Cub cylinder heads, pistons, flywheels, crankcases, differential cases, rear axle housings, steering gear housings, and transmission cases. The McCormick Works was the source of the Cub steering gear housing bases and many malleable iron castings. IH's Canton Works and Rock Falls Works both made "scaled down" implements to be used with the tractor that "worked like a bear, but was a Cub in size."

1950s Refinement of the Cub

The 1950s would see IH expand the Cub tractor line with a new "low-slung" model called the Cub Lo-Boy. The Cub Lo-Boy was introduced to meet the growing sales in the industrial and commercial markets. New advancements in hydraulics and basic styling changes would dominate the Cub line in the 1950s. IH built four differently styled Cubs in the 1950s, more than at any other time in the tractor's history. After an initial surge in sales, the Cub sales tapered off. IH had several ways to help bolster sagging sales and retain or grow market share. The Mid-Century white demo tractor program was one of these programs. At IH, it was time to make tractors and get Cubs out into the hands of potential users.

FARMALL CUB MOVING SNOW
A Farmall Cub with front-mounted grader blade quickly clears the snow from the station platform at the Wayne, Illinois, train depot. The daily express was due and with it the mail and other cargo that would be handled by the Cub. *Wisconsin Historical Society*

FARMALL CUB WITH MOUNTED PLOW

Even when equipped with a light, one-button moldboard plow, the Farmall Cub did not set any plowing speed records. Thankfully, the tractor's target audience didn't farm massive acreage. Even though the Cub had only 9 horsepower, the low gearing in the Cub's transmission made this tractor a relative powerhouse in the field. *Wisconsin Historical Society*

CUB TRACTOR WITH AUGER DRILL

Drilling holes with an earth auger was never easier than when you had a power auger mounted on the Cub. If you needed to install a fence or plant trees or shrubs, this was the easy way to accomplish the task. *Wisconsin Historical Society*

Closure of the Louisville Plant and Foundry

The 1970s would mark the twilight hours for the Cub and Cub Lo-Boy tractors. Increasing market competition in the compact tractor segment coupled with an apparent lack of research and development (R&D) by IH, and continually diminishing sales would bring the thirty-plus-year production run of the Cub to an end. The closure of IH's Louisville plant and subsequent sale of the Cub Cadet line to the newly formed Cub Cadet Corporation would be the outcome.

After more than thirty-two years in production and two major tooling changes (the machine tools and molds simply wore out, twice!), it was obvious that the Cub would need to be retired. To rebuild the Cub tooling for a third time would be cost ineffective. The three major factors behind this

184 TRACTOR WITH MOWER
The 184 Cub Lo-Boy tractor (shown here fitted with a 60-inch-cut mower deck) would be the last of the Cub Lo-Boy series tractors made at Louisville Works. When the deal to sell the Cub Cadet tractor line to MTD Products Inc. of Cleveland Ohio was made, the 184 Cub Lo-Boy was not included and the tractor was retired. *Wisconsin Historical Society*

decision were cost of production (new tooling needed to be made, again!), an aging production facility with a high manufacturing cost, and new engine emission regulations that were scheduled to take effect in the next decade. All three of these points plus a few others would seal the fate of the beloved Cub. Tractor manufacturers were turning to compact diesel-powered tractors (usually built in Japan) for a fuel-efficient, reliable, and clean-burning power source. Advancements in manufacturing and hydraulics had outstripped the Cub's design and marketability. IH's replacement for the Cub was a diesel-powered compact-size tractor, built in Japan.

The final production Cub tractor was built in March of 1979. In a notarized letter dated February 18, 1980, IH notified Miller True Value Hardware of Charlotte, North Carolina, that it was the recipient of the Final Production Cub Tractor. The letter states: "The International Harvester Company Louisville Plant began production of the Cub model in 1947. The customer acceptance of this product resulted in an amazing production record, which continued until March of 1979. During the nearly thirty-three years this model was manufactured at Louisville, 245,831 Cub tractors were built."

The letter continues: "We are understandably proud of quality products produced at the Louisville Plant which have served the American Farmer and have contributed to his achievements. The end of an era for the Cub tractor was reached with serial number 253685. We congratulate you for being a part of this success story as the recipient of the last Cub tractor."

A friend of this author contacted the Miller family who stated that they had kept the final built Cub, never selling it to anyone. Unfortunately, the tractor was lost to a tragic shed fire a number of years ago. The family stated that the tractor was not repaired. The disposition of what remained of the tractor is unknown to this author. One would surmise that it was scrapped.

There is one point that has often been dismissed as not being true. This is that the last Cubs made were actually painted entirely red, not yellow and white as they had previously been. The final Cub that was built was painted red and had a long, blue stripe on the side of the hood. The blue stripe was the same design used on yellow-and-white version models.

It is known that the Farmall Cub was the longest-running production tractor of all time. IH's decision to close Louisville in 1980 was influenced by many factors. Dresser Industries was sold, reducing the need for the plant. After thirty-two years of tractor production, the factory was not cost efficient to run anymore. The drastic downturn in the North American ag economy in the early 1980s coupled with poor management decisions and continued top-heavy mismanagement led to huge financial loses by IH. Cash for physical plant revival was nonexistent. The Cub finally went into a permanent "hibernation." IH had planned to build components of its new TR4 and TR3 farm tractors at Louisville, but the economy and tractor sales both soured, adding yet another nail into Louisville's coffin.

IH closed Louisville and transferred the Cub Cadet production line to its new facility in Brownsville, Tennessee. The foundry closed and the plant was eventually scrapped out and later razed. Today, the area that *was* the former Louisville Works is now part of the Louisville Airport. This airport is most widely known for it being the main air freight terminal for United Parcel Service (UPS). How ironic it is that on the same ground where Cubs, Super As, 140s, and Cub Cadets were once built, their repair parts (packaged in boxes) are directed toward a servicing dealer to keep a tractor running.

CHAPTER 2

Cub Model Guide

Original 1947–54 Cub

The Farmall Cub that hit the dealers' lots in 1947 remained relatively unchanged until 1955 when IH restyled the tractor. The first Cub tractors did not have the IH hydraulic lifting system fitted on them. This was called the "Touch Control" system by IH, and it was a system that lifted implements with just an easy touch of the control. Touch Control was not ready for production when the Cub was first released to the market. IH had the system ready, however, not long after the Cub introduction, and some early 1947–1948 Cub tractors were sold equipped with the Touch Control system. Except for some minor decal changes on the hood (mainly the deletion of the Deering name and the Circle Cub decal) the Cub remained unchanged until 1955.

The first production Cub (SN# 501) was set on the assembly line at Louisville Works to start its journey into history on Saturday afternoon May 10, 1947. At the end of Monday the 12th, the Cub was barely halfway down the assembly line. Tuesday brought more progress to the "pilot run," or assembly test model, as it was starting to look like a tractor. At 1:50 p.m. it was out of the paint booth and drying. It reached its assembly culmination, the end of the line, at 2:40 p.m. Louisville Works Manager J. E. Harris boarded the first Cub and at 3:05 p.m. started it up. This also marked the start of the longest production run of a single model tractor in IH's company history.

IH sent Cub #501 to the Canton Works for test fitting of the many implements that were to be offered with the Cub. The next five Cubs, serial numbers 502, 503, 504, 505, and 506, were outfitted with Spanish language decals and sent to the IH Works in Saltillo, Mexico. Farmall Cub serial number 501 was eventually retailed to a farmer in Wisconsin. This farmer also purchased nearly every single Cub

REMOVING CORN TASSELS
This Farmall Cub tractor was modified for removing corn tassels by Knudsen Implement Co. of Manning, Iowa. The tractor was owned and operated by Gruhn Hybrid Corn of Manilla, Iowa, and is shown in 1949. *Wisconsin Historical Society*

FIRST CUB TRACTOR BUILT

With Louisville plant manager J.E. Harris at the controls, the first Cub tractor built was driven off the assembly line at 2:40 p.m. on Tuesday, May 13, 1947. This tractor (serial number 501) was sent to IH's Canton, Illinois, factory to test-fit Cub-specific implements. *Wisconsin Historical Society*

FARMALL CUB WITH REEL GANG MOWER

Fitted with an under-mounted reel-type mower, this Cub tractor is making quick work of the yard. The mower is lifted by linkage connected to the hydraulic Touch Control lifting unit. *Wisconsin Historical Society*

FARMALL CUB WITH CULTIVATOR
Cultivating valuable vegetable crops, the Cub with mid-mounted cultivator excelled. The offset operator's station offered unmatched forward visibility that IH called "Culti-Vision." *Wisconsin Historical Society*

FARMALL CUB WITH FIELD CULTIVATOR
This Farmall Cub has a rear-mounted field cultivator fitted to it. The long lever to the right of the driver is the depth control for a plow, and is not used with this cultivator. The two chains allow the unit to flex with the contours of the field. *Wisconsin Historical Society*

FARMALL CUB WITH FERTILIZER AND PLANTER ATTACHMENTS

Fitted with both a fertilizer applicator on the tractor and a rear-towed planter, this Cub is busy planting a crop for its owner. The bright white color in the black-and-white image is dirt or dust, not white paint. *Wisconsin Historical Society*

FARMALL CUB WITH DISC HARROW

With a disc harrow in tow and factory installed headlights, this Farmall Cub is ready to work all day (and night) if needed to finish the fieldwork. *Wisconsin Historical Society*

FARMALL CUB ON DISPLAY TRACK

IH needed to promote the sale of the new Cub tractor, and a great way to do this was to have a circle-driving course set up at fairs and other exhibitions. In this image, a man who appears to be entertainer Arthur Godfrey is driving the Cub. *Wisconsin Historical Society*

FARMALL CUB CULTIVATING AND APPLYING FERTILIZER

Cultivation of new crop plants was easy with the Cub Culti-Vision feature. The large tank carried granular fertilizer that was applied while cultivating weeds, allowing two tasks to be done at one time. *Wisconsin Historical Society*

implement available at the time for use on his garden produce farm. Only recently has this Cub been restored to its original condition. The locations of Cubs 502 through 506 are currently unknown to this author; they are believed to have been scrapped.

The list price of the Farmall Cub with standard equipment was $545 F.O.B. Louisville Works in 1947.

1950 White Demonstrator Cub

When the Cub was first introduced in 1947, sales skyrocketed for the first two years, then they leveled off and even started to decline in numbers. IH marketing was quick to recognize this trend and launched a massive tractor demonstration program that would not only mark the middle of the twentieth century (1950), but also IH's dominance in the marketplace since its formation in 1902. IH hoped that the sales volume of the Cub and the other Louisville built tractors would rebound.

Demonstration tractors are machines that were lent out to farmers; and every once in a while, IH elected to make demonstrator models with unique markings that could include paint. The Mid-Century 1950 program was to paint three of their IH models white. The bright white sheetmetal was dressed up with gold star decals. Large cardboard placards were used to indicate interesting features (and a few collectors have not only the white machines, but these cardboard bits on them to really take you back in time). These rare and highly desirable machines are known as the "White Demo" model tractors. The Farmall Super A, Farmall C, and Farmall Cub are believed to be the only three models offered as white demo models. All three were built at the IH plant, Louisville Works.

WHITE DEMO CUB WITH SICKLE MOWER
This 1950 white demonstrator Cub is fitted with a side-mounted sickle bar mower. The white cutter bar is incorrectly painted, but the blue guards attached to it are painted the right color. IH wanted dealers to demonstrate tractors with implements to show the versatility of the Cub. *Lee Klancher*

1950 WHITE DEMO CUB
In 1950, IH painted several thousand Cub tractors white for their Mid-Century tractor demo program. Dealers were to use the special white-painted tractor for demonstration purposes to encourage customers to buy the Cub tractor and its implements.
Lee Klancher

WHITE DEMO CUB
Close-up shot of the IH-built C-60 engine used in the Cub tractor. The carburetor was also built by IH. The oil-bath-style air cleaner is between the engine and radiator.
Lee Klancher

WHITE DEMO CUB

The implements that were lifted hydraulically by the IH Touch Control lift system attached to the lifting rockshaft arm shown here.
Lee Klancher

To bring attention to the IH tractor line and draw the attention of potential customers, Louisville Works would take a few weeks and produce only white-painted tractors. Once the run was done, they would switch back to doing red tractors. Several runs of white tractors were produced in 1950 during the months of January, February, and March.

Only the Farmall Super A, Farmall C, and Farmall Cub are believed to be the true white demonstrator tractors that IH offered as factory-built models.

IH used an aggressive marketing program to get demonstrations and (hopefully) sales of IH tractors and machinery rolling. IH encouraged its dealer network to outwork, outsell, and out-demonstrate the IH line to not only loyal IH customers, but to those who had Brand X or were new to the marketplace. Large field demonstrations that had a "county fair" look to them were held nationwide in the spring and summer months of 1950. To attract potential customers, dealers used colorful tents, placed the white machines in tractor parades, and adorned the machines with decorative bunting. IH encouraged dealers to place the Cub tractor with Boy Scout of America troops, who often accompanied the tractors in parades or displays. Some dealers even assigned their salesmen to travel to area farms or commercial markets to demonstrate to prospective customers the Cub system of farming.

Once the demonstrations were done, IH instructed dealers to repaint the tractors red. Some tractors were sold and left the dealership with white paint—how many is not known.

Authoritatively identifying an original white demonstrator tractor is difficult, as the serial number record of which tractors were painted white no longer exists.

One key is that all the white-painted models were built at the Louisville Plant. Tractors built at this plant have a cast letter "L" on the clutch tube housing, which is typically upside-down when compared to the other numbers cast into the part.

Another method to determine if a tractor was a white demo is to scratch the underside of the hood or other hidden area that was commonly missed by the process of repainting. Under the red paint, you can typically see the original white paint. When the white demo tractors were repainted to IH red, very few were completely disassembled to the point that the factory did when it painted them.

The most obvious method of verifying a true white demo from a fake is to examine the tractor's serial number. The popular belief is that the white Farmall Cubs were created from some of the machines between serial number 99356 to 106516. So not all of these Cubs were white, but if the serial number is outside those numbers, it most likely was not painted white originally.

IH encouraged its dealers to sell the Cub system of farming to farmers. One of the big advantages IH had on comparable small tractors from other manufacturers was the large line of IH implements engineered specifically for the Cub. When buying a Cub implement, the farmer could be assured that the implement he bought was sized right for the Cub. Not too big, not too small.

1950 Cub Standard Equipment List

A standard-issue Farmall Cub came equipped as follows in 1950:

Four-cylinder C-60 gasoline engine
Vertically adjustable draw bar
Adjustable rear wheel tread
Fenders
Foot-operated differential steering brakes
Three-speed transmission
Magneto ignition
Non-adjustable front axle
Spring-mounted implement style seat
3.00-12 inch 2-ply front tires
6-24 inch 2-ply rear tires
Farmall Cub price with
    standard equipment $659

Cub Optional Equipment

A list of optional equipment could be ordered with your Cub to tailor it to your needs.

Adjustable tread front axle	$21.75
Belt pulley and power take-off attachment	$37.00
Swinging draw bar	$4.50
Electric starting and lighting	$59.50
Combination rear lamp and tail lamp	$4.50
Spark arrestor	$6.00
Muffler	$2.75
Deluxe foam rubber upholstered seat	$8.25
Touch Control hydraulic system	$84.00
Front-wheel weights	$6.00 per pair
Rear-wheel weights	$30.25 per pair

A factory standard Cub came equipped with electric starting and lighting, Touch Control hydraulics, wheel weights (one set front and rear), muffler, upholstered seat, adjustable front axle, and belt pulley/power take-off. The retail price for this package was $867.

That was a lot of money in 1950, but it pales in comparison as to the prices that Cub tractors can bring today in the marketplace. Sale prices of Farmall Cubs today can easily exceed triple that of the original price for a completely restored tractor in the collector marketplace.

WHITE DEMO CUB
The tractor's IH-built magneto ignition is on the right-hand side of the C-60 engine. Many Cubs were fitted with magneto ignition. The electric starting motor built by Delco Remy is in the lower left of the photo. *Lee Klancher*

WHITE DEMO CUB
The manual steering gear box was also part of the front bolster. This casting also served as the lower radiator tank and was bolted to the implement mounting base. The square hole in the casting is where the implement attached. *Lee Klancher*

1950 Cub Implement Price List

To go along with your new Cub, you needed Cub-specific implements. Using implements that were designed specifically for the Cub meant that more work could be done in less time than if a larger "oversized" implement were used. IH offered a vast array of implements to fit your specific crop needs. A short list of these implements and their 1950 list prices include:

#3 spring-tooth field cultivator	$61
#23A 4-foot-wide tandem disc harrow	$134
Cub 33 two-row bean harvester	$55.50
Cub 22 4 1/2-foot-wide sickle mower	$97
Cub 151 one-furrow disc plow	$128
Cub 189 one-furrow 12-inch two-way moldboard plow	$118
Cub 54 leveling and grader blade	$35
Cub two-wheeled farm trailer with tires	$174
Cub rear-mounted tool bar	$28.25 (ground tooling available at extra cost)

IH dealers often would deal in tractor and implement packages where if you purchased the tractor and plow, for example, the plow would be sold for less than list price or there would be some other financial incentive to sweeten the deal and secure the sale.

Cub Lo-Boy

The Cub Lo-Boy was a variation of the Farmall Cub tractor. The low-slung Farmall Cub was first shown at the Hinsdale, Illinois, demonstration show held September 29 and 30, 1954. Farm Tractor Committee Report Number 370, dated December 21, 1954, traces the development of the new low-slung Farmall Cub that was referred to as the Cub Lo-Boy.

A stock Farmall Cub was the base unit used to build the prototype of the low-slung tractor. Rotating the rear axle housings modified this Cub, thus lowering the tractor height by about 7 inches. The front axle extensions were also shortened accordingly to match. With the rear axle housing rotated forward, the tractor's wheelbase was shortened to 62.5 inches as compared to the 69.25-inch wheelbase of the Farmall Cub.

The other modifications made to the prototype included moving the operator's seat to an area where it was nearly inline with the steering wheel. The seat would be made of sheet metal construction with foam rubber padding that would have a tilt-back feature. This was in contrast to the post-and-spring style seat on the Farmall Cub. The low-slung Cub could be mounted/dismounted from either the front or rear. The transmission shift lever was also shortened and bent forward slightly for better operator control.

Due to the acceptance of the prototype by those who viewed it at the Hinsdale show, four more units were modified accordingly by the Farm Tractor Engineering Department and were sent to regional farm shows to gather more market-interest data. Would the new low-slung Cub hinder the Farmall Cub

CUB LO-BOY TRACTOR
The Cub Lo-Boy was the low-profile variant of the Cub tractor. The lower center of gravity made the tractor more stable for many tasks. It was also easier to get on and off the tractor, due to the lower height. *Lee Klancher*

sales? How much new market share could IH gain by introducing this model? Were there any competitive designs that this could be directly marketed against? Would sales of the "regular" Cub be reduced? These were all questions that IH needed to answer before mass production would begin.

Of the five prototypes built, the Farm Tractor Engineering Department kept two, and three were sent to the Implement Works. It was imperative that the Implement Works had implements ready to be used with this tractor when it was introduced for sale to the public.

With the low-slung Cub, accessibility to drainage ditches would be vastly improved over the Farmall Cub, due to the lower center of gravity point. IH determined that most of the Cub Lo-Boys made would be sold primarily for mowing jobs. It was imperative that the Allied Equipment manufacturers had a lawn-mowing attachment ready for the tractor introduction.

The uses for the new Lo-Boy were nearly as endless as those tasks that the Farmall Cub could perform. Also, because the Cub Lo-Boy was felt by IH marketing to be used primarily by industrial/municipal applications, an electric horn was to be made as an available attachment. The Cub Lo-Boy was typically equipped with an under-slung exhaust system but a few were equipped with a vertical exhaust. A vertical exhaust defeated the purpose of making a lowered tractor. The basic components of

CUB LO-BOY FAST HITCH
IH's new Fast Hitch attachment was also fitted on the Cub Lo-Boy tractor. Here, a single hitch socket accepted a special hitch tong on the implement. By backing into the implement and raising the hitch, the connection was made. *Lee Klancher*

CUB LO-BOY
The Cub Lo-Boy has an underslung (horizontal) exhaust system that exits the rear of the tractor. This helps to keep the overall profile of the tractor low. Dual sealed-beam headlights are also found on this tractor. *Lee Klancher*

the tractor—engine, transmission, differential, and cowling—were common to both the regular Cub and the Cub Lo-Boy.

The General Sales Department forecasted that if the Cub Lo-Boy were produced, more than 2,500 units could easily be sold in 1955. This was in addition to the already projected sale of over 10,000 Farmall Cub tractors for 1955. The report stated last that the name of this new tractor should be "International Cub Lo-Boy." The "International" name came about as a new indicator of standard-tread tractors like the International 600 and 650. These "big brothers" to the Cub Lo-Boy offered a fixed or very limited adjustment of the tractor's wheel tread. The list price of the Cub Lo-Boy in 1955 was $980. This is a fraction of the amount that these tractors command today in the marketplace.

Fast Hitch
The addition of the rear Fast Hitch allowed the (1955 and later-built) Cub and Cub Lo-Boy to be used with Fast Hitch implements such as the moldboard plow, disc harrow, and rear platform carrier attachments. The Cub Fast Hitch differed from the larger IH tractor's Fast Hitch in that it only had a single hitch tong receptacle centrally located at the rear of the tractor. Cub Lo-Boys and Cub tractors equipped with a Fast Hitch were at the time referred to as one-point hitch tractors. Most larger Farmall tractors had two hitch tong receptacles on them. These tractors equipped with a Fast Hitch were often referred to as two-point hitch tractors. The implements that fit into these sockets have a specific thickness of hitching point or tong.

With a Fast Hitch the operator could back up to the implement, "click" the implement hitch tong into the tractor hitch socket, and go. You never left the seat. Changing implements was never faster, or easier! The only drawback to the Fast Hitch system was the lack of hitch draft control. IH's main competitor (in implement hitching systems) was the Ferguson system. This was found on the wildly popular Ford 2N, 9N, and 8N tractors. It offered a crude draft control system that would automatically adjust the hitch to a preset level made by the operator. If IH engineers could devise a draft control system into the Fast Hitch, it could become the industry standard of implement-to-tractor attachment. IH refused to release the Fast Hitch patents to the rest of the industry, and when the Ferguson patents expired, most tractor manufacturers adopted this system. Allis-Chalmers suffered a similar fate with their "snap coupler" rear hitching system later in the 1960s, too!

When the Cub Lo-Boy was introduced, IH started production with serial number 501. IH kept the Lo-Boy serial numbers recorded separately from the traditional Farmall Cub serial number list. Whenever IH changed the body style of the Cub and Cub Lo-Boy, the serial numbers did not start over at 501 again. The Cub/Cub Lo-Boy service parts book points out the serial number breaks for the respective tractor body styles.

The Cub Lo-Boy was never offered with a "wire mesh" style front grille like the original Farmall Cub (built from 1947 to 1954) was. The "00" and "50" series styled bar grilles were the only styles offered on the 1950s-built Cub Lo-Boy. All of the first two generations of Cub Lo-Boys were equipped with stainless steel emblems on their hoods, with the addition of white-painted grille and white decals on the side of the hood—as well as white rear wheel centers and white front rims—being the major difference between the first and second generation of

CUB LO-BOY TRACTOR WITH MODEL 100 SPREADER
IH Cub Lo-Boy tractor with McCormick model 100 manure spreader. The model 100 spreader was first built in 1941 and sized appropriately for the Cub and Cub Lo-Boy tractors. *Lee Klancher*

CUB LO-BOY ENGINE DETAIL
Right-hand view of the Cub Lo-Boy grille and emblems, showing the stamped aluminum emblems that IH used on this vintage of tractors. These emblems today are rare and can be quite valuable (and difficult) to find! *Lee Klancher*

Cub Lo-Boys. The third generation of Cub Lo-Boy used stamped-metal, oval-shaped plates for model identification and had a large four-horizontal-bar front grille. These were similar in appearance to the 460–560 styled tractors of the same era. IH gave all of its Cubs and Cub Lo-Boys built through 1963 styling that was similar to the larger Farmall tractors being produced at the same time for "family appearance" reasons. The Cub Lo-Boy's standard color was changed from IH Red to 483 Federal Yellow in 1960.

The Cub was first released in 1947 with a wire mesh grille and smooth-sided hood having three horizontal bars stamped into the hood side for rigidity. This body style was retained until the 1954 tractor production year. At this time, the Cub and the newly introduced Cub Lo-Boy shared the same hood, having a new nine-horizontal-bar grille added. This hood had a wedge-shaped depression stamped into its side directly above the engine. The tractors were built from 1954 to 1958. IH made two variations of these styled tractors.

All Cubs made from serial number 185000 to 210000 and Cub Lo-Boys from serial number 501 to 10000 have stainless steel emblems attached to the side of the hood identifying which model tractor it is. After 1956, the indentation on the side of the tractor hood had a white decal added that was a background for the letters. The reason for this was to emulate the red/white paint scheme IH was using on its big tractors of the time: the 350 and 450. These "big" tractors had a two-tone paint job, and the Cub and Cub Lo-Boy were always styled to match the other IH farm tractors being built at the time for family appearance. Much confusion exists about the 1954–1959 styled "stainless" Cub tractor family, with incorrect information appearing on discussion lists, in books, and in the minds of countless collectors. The largest falsehood that is continually perpetuated is that

CUB LO-BOY OPERATOR'S PLATFORM

The square cushion seat of the Cub Lo-Boy has the tractor's battery stored underneath it. The heavy padding of the cushions provides a smooth ride for the operator. The controls in the image are for a Model 1000 loader. *Lee Klancher*

CUB LO-BOY REAR HITCH/PTO
The rear PTO shaft extends from the transmission housing on the Cub Lo-Boy. The shaft has six splines like a 540 PTO drive, but is smaller in diameter and turns at engine operating speed. The rear work light is also in the photo. *Lee Klancher*

these tractors had model identification decals. This is not true. The only decals on the stainless tractors are the white hood background (used on the 1957–1959 built tractors) and the various caution/warning decals found on all of the machines. At one time, IH did offer the stainless emblems in decal form through their service parts division, as replacement service parts, but these parts have since been retired.

One important distinguishing feature that all of the Cub and Cub Lo-Boy tractors have is a two-wire sealed beam headlight in these machines. To distinguish the sealed beam lights from the older style, one must look at both the wiring and the shape of the light housing. The sealed beam lights have two wires leading to them and their metal housings have a flat profile to them. This author calls them the "pie pan" lights as the housings have the shape of miniature pie pans. The older style lights are often referred to as the "bullet" style lights due to the unique bullet-shaped housing they are contained in. Both styles of lights were a product of the GUIDE Light Company that is now owned by the AC-Delco division of GM. All of the headlights and taillights offered by IH were painted to match the tractor. IH did not use black-painted lights!

Another electrical note is that all Cubs prior to serial number 224401 and Cub Lo-Boys prior to serial number 18701 were made with a 6-volt *positive* ground electrical system. In about mid-1964, IH "modernized" to a 12-volt negative ground system. Caution must be exercised when jump-starting or charging a 6-volt system. The battery must be correctly installed with the positive cable going to ground on the tractor chassis. Failure to heed proper polarity can and will result in machine damage and/or your own injury or death! A common electrical-related question is "What are those two holes in the hood of my Cub on the right side?" They are

FAST HITCH ON CUB LO-BOY
This Cub Lo-Boy is fitted with an aftermarket front loader, but it also has IH's very own rear Fast Hitch, which features a socket mounted to the tractor's rear lift arm. The Fast Hitch implement has a tong that engages the socket to connect the implement and tractor together. The operator never has to leave the seat to couple or uncouple the implement. *Wisconsin Historical Society*

access holes to the Cub's generator bearings. Since the generator is tucked under the hood, and the hood does not have a flip-open access door, IH engineers added the two holes to allow the operator using an oilcan to add two or three drops of 20W oil to the bearing caps of the generator.

Today, to find a Cub or Cub Lo-Boy with the stainless emblems intact can be quite a find. Reproduction emblems are being marketed now, but at a high price. The average list price of a 1950s Cub Lo-Boy was $1,100. Today, a Cub Lo-Boy in good mechanical and physical condition can bring three or four times that original amount. Maybe it would have been a good investment to buy new tractors in the 1950s, store them for thiry or forty years, and resell them for huge profits?

1955–58 Cub

Beginning with serial number 185001 and first produced in 1954 as a 1955 model, the Cub styling was updated to match the Hundred Series tractors, led by the 300 and 400 tractors. The Cub received a new grille that was similar in design to the larger tractor models. IH also introduced a new low-profile model of the Cub tractor called the Cub Lo-Boy.

Both Cub models also featured a new IH innovation available as an option; a quick-connect hitch that IH called the Fast Hitch. The Fast Hitch on the Cub and Cub Lo-Boy was unique in that it had only one mounting prong/socket for the hitch instead of the two found on the larger tractor models.

In 1956, the styling changed a bit as the grille and side panel on the hood were painted white. The serial number break for this change was not available at press time. The tractor was mechanically the same as the previous model year Cub. The Cub Lo-Boy final drives were rotated to lower the rear of the tractor and the front axle was lowered as well.

FARMALL CUB WITH DIGGER
The newly restyled Cub tractor used stainless steel emblems instead of decals for the model labels on the tractor. This tractor has the grille painted white; it should be painted red to be factory-correct.
Lee Klancher

CUB ENGINE DETAIL

The IH-built, four-cylinder C-60 engine on the Cub was a simple flat-head design. IH engineers designed the Cub so that the gas tank/hood did not need to be removed for basic maintenance. *Lee Klancher*

CUB AT WORK

The short gas tank cap on this Cub is an original IH item. Many years later, IH was sued over an incident involving a gas cap on a tractor and a new, taller, triple-baffle cap was released for service. This tall gas cap replaced the flat gas cap. *Wisconsin Historical Society*

CUB DASHBOARD
The dash panel on the Cub is very basic, with an ammeter, a push-pull, on/off ignition switch, and a rotary light switch knob. The Cub was a basic, no-frills tractor that was easy to operate and repair. *Lee Klancher*

CUB LO-BOY WITH FLAIL MOWER
The Cub Lo-Boy was styled to match the Farmall Cub. The Lo-Boy looked like a little brother to the Cub due to its lower profile, but was dimensionally nearly the same otherwise. *Lee Klancher*

CUB LO-BOY GRILLE DETAILS

The grille screen on the Cub and Cub Lo-Boy is fastened with screws that allow for its quick removal. This lets you clean the radiator easily. The Lo-Boy and Cub emblems are made of stamped stainless steel for durability and appearance. IH switched to this style of tractor decoration in the mid-1950s.
Lee Klancher

CUB FRONT-WHEEL WEIGHT
The cast iron front-wheel weight of the Cub and Cub Lo-Boy not only assisted in tractor stability, but was later used as the rear-wheel weight in the IH-built Cub Cadet garden tractor. *Lee Klancher*

CUB LO-BOY EMBLEM
The Cub Lo-Boy serial number plate is located on the right-hand side. *Lee Klancher*

1959–63 Cub

The 1960s saw IH change the Cub and Cub Lo-Boy line like no other time in its history. Not only were new body styles created, but a new design of Lo-Boy tractor was engineered that was radically different from the original versions. The Cub, too, would undergo change with the Farmall version being retired, but it was the Lo-Boy that had major reengineering done.

The first major change came for the 1959 model year, when the Cub and Cub Lo-Boy were restyled to look like the new IH 460 and 560 models. These tractors had a large four-horizontal-bar grille that was painted IH white, along with the tractor's model designation in a stamped metal oval located on the side supports of the front grille housing. The tractors were still bathed in IH red, but now they were accented in IH white. The Cubs from serial number 210001 to 222500 were styled like this, as were the Cub Lo-Boys from serial number 1001 to 17200.

On March 31, 1960, IH decided to replace the familiar red paint it had been using on the Cub Lo-Boy with 483 Federal Yellow. The Farmall Cub retained IH red as the standard color, and 483 Yellow as an optional color choice.

FARMALL CUB WITH SICKLE BAR MOWER

The Farmall Cub, when fitted with a side-mounted sickle bar mower, can cut acres of hay or grass in a day. The offset operator's platform offers an unequaled view of the mower working. *Wisconsin Historical Society*

FARMALL CUB WITH MOLDBOARD PLOW

When at the headland of the field, the operator simply moves a lever to raise the moldboard plow. Once turned around, another touch of the lever lowers the plow into the ground. *Wisconsin Historical Society*

CUB WITH CULTIVATOR
A backyard garden is no match for a Farmall Cub and cultivator. The Culti-Vison view from the operator's seat is unmatched by the competition. The twin sealed-beam headlights let you work late at night. *Wisconsin Historical Society*

1963 CUB WITH TWINDRAULIC LOADER
A survivor tractor-and-loader combination shows how both the tractor and loader were when they were new with original decals. Finding "survivor" units like this today can be challenging. *Lee Klancher*

CUB ENGINE

A close-up view of the engine on the Cub showing the carburetor and horizontal exhaust system. The carburetor has "IHC" cast into the bowl, meaning it is an original IH-built carb. *Lee Klancher*

CUB WITH CULTIVATOR

The deluxe silver seat on the Cub was designed with operator comfort in mind. The backrest cushion support with armrests made the task of cultivating long hours on end less tiring to the operator. *Wisconsin Historical Society*

CONTROL LAYOUT ON CUB WITH 1000 LOADER

The Cub's operator's dashboard has an ammeter to monitor the electrical system. The silver button switch is for the tractor's ignition. Pull out to run, push in to stop. *Lee Klancher*

FARMALL CUB EMBLEM

The 1959 to 1963 Cub tractor has a stamped aluminum emblem badge. The Cub badge is painted two colors (red and white) to match the rest of the tractor. Finding these emblems today can be quite challenging. *Lee Klancher*

1964–75 Cub

The Farmall Cubs received their last styling facelift in 1964 with a new flat grille housing being added. Cubs of serial number 222501 and higher have this "square front" styling. The grille screen, hood, front wheels and rear discs, and grille housing were now bathed in IH white with the balance of the tractor painted IH 483 Federal Yellow. This

flat-grille Cub body style would be retained by IH (with some minor color and decal changes) until the Cub was retired in 1979. The last Cub Lo-Boy with the square grille was serial number 26007 built in 1968. The Model 154 Cub Lo-Boy replaced this that same year.

As IH noted with the release of Tractor Committee Report #141 dated May 25, 1964, the increase in typical acreage on the farm at that time eroded Cub agricultural sales. In this report it was stated that when the Cub was originally introduced, it was primarily used on small farms. By the mid-1960s, most full-time farmers were purchasing large tractors to work more acres with fewer people. These trends meant the agricultural sales of Cubs steadily declined during this time. The upside of this story is that industrial use of Cubs increased steadily, and by the mid-1960s accounted for 75 percent of the annual Cub tractor sales.

Since the majority of Cub sales were being made to industrial users, the International version of the Cub was to be retained for production and the Farmall (agricultural) version would be dropped. IH offered the International version of the Cub in its Federal Yellow and IH White paint scheme. It was not felt that this color change would hinder any sales to ag customers, nor would overseas sales be affected. So the red Farmall Cub was retired from the IH tractor line for the first time. This change took place in 1963. The International Cub was built from serial number 224704 starting in 1964 to serial number 248125, which ended the 1975 year production. The initial list price of the International Cub in 1964 was $1,680; while in 1979 (when it was dropped from the IH tractor line) that list price had doubled to $3,529. It is interesting to note that while a vertical exhaust system was available on the International Cub, the IH sales literature of the time never showed this feature.

1964 CUB WITH BLADE

The Cub was restyled in 1964 to have a new flat grille housing. These tractors are often referred to as the flat-face Cub. IH would later update the grille housing in the 1970s, but the flat-face design was carried through until the Cub ended production. *Lee Klancher*

Industrial Cub

IH sales of the Cub and Cub Lo-Boy tractors in the 1960s were being directed and sold to commercial and industrial users such as landscapers, groundskeepers, and small industry (foundry, etc.), and a large number were sold to city, municipal, state, and other governmental departments and entities. IH recognized this and added the Cub tractor to its industrial line of equipment. This tractor was termed the "Cub Industrial" tractor and was identical to the Cub tractor offered for sale by the ag group except in name (and possibly price) only.

The tractors' decals were not different, and the word "Industrial" was not added to the serial tag or to the decal stripe. IH sold thousands of Cub tractors for industrial/commercial use with the two most common applications being for mowing of parks, cemeteries, and roadsides, and for general grounds

YELLOW CUB LO-BOY AT IH INDUSTRIAL DISPLAY
The yellow Cub Lo-Boy (second tractor in front row) is part of an IH Industrial Equipment display at a local event. IH painted the industrial tractors Federal Yellow (the standard color) or other colors, like Omaha Orange, to meet the customers' needs. *Wisconsin Historical Society*

CUB LO-BOY WITH DANCO SNOW BLOWER
After a successful demonstration, workers are loading this Cub Lo-Boy fitted with a Danco-brand front-mounted snow blower back onto the delivery truck. On-site demonstrations often lead to many tractor sales. *Wisconsin Historical Society*

CUB LO-BOY WITH CABLE TRENCHER

A rear-mounted cable trencher is installing utility cable. The trencher is powered by the tractor's PTO. The small size of the Cub Lo-Boy made it ideal to work in tight places. *Wisconsin Historical Society*

INDUSTRIAL LO-BOYS

These Omaha Orange-painted Cub Lo-Boys fitted with side-mounted sickle bar mowers were being stored by the Minnesota Department of Transportation (DOT) while they awaited work on the roadways. The dual rear wheels on the Cub Lo-Boy added stability when cutting roadside weeds. *Wisconsin Historical Society*

CUB MODEL GUIDE

CUB LO-BOY WITH FLAIL MOWER

Fitted with dual rear wheels for added tractor stability, the Cub Lo-Boy also has a flail-type mower added. A large spinning drum on the mower has flair knives that cut the grass when the drum spins. *Wisconsin Historical Society*

CUB WITH FORKLIFT

Piper & Paine Manufacturing offered a forklift attachment that converted a front-driving Cub tractor into a rear-driving forklift. Modifications to the tractor's steering, seat, and controls were just some of the changes needed. *Lee Klancher*

CUB FORKLIFT
Close-up view of the steering changes on the Piper & Paine Cub forklift attachment. Notice the roller chain to connect the steering wheel to the control shaft. *Lee Klancher*

CUB FORKLIFT
Operator's platform on the Cub with Piper & Paine forklift. The reversed seat, hydraulic reservoir, and control valve are by the operator's seat. The Cub gearshift lever location remains unchanged. *Lee Klancher*

CUB FORKLIFT
The rear view of the Cub fitted with Piper & Paine forklift. The operator's view is in line with the lifting mast of the forklift. The large box underneath is for counterbalance weights. *Lee Klancher*

keeping at schools, universities, or other such public institutions. The exact number of Cub Industrial tractors would be difficult to obtain as it has no unique features other than the name IH used to sell it by.

The Cub Industrial was dropped from the IH Industrial equipment line by 1972.

Since IH was a worldwide corporation, its products were made in one country for distribution in several others. The Cub and Cub Lo-Boy were products that shared this distinction. The Cubs and Cub Lo-Boys sold in the North American market had decal insignia on them stating "McCormick Farmall Cub" or "International Cub," while an export tractor might have "McCormick International Farmall Cub" on its hood. The reasoning was that those sold in offshore markets still had registered trade names in their respective countries of sale that needed to be used. It's amazing that with all of the trademark names that IH had registered around the world that a label with all of these trade names even fit on the side of the Cub's short hood.

Model 154 Cub Lo-Boy

IH made a major change to the Cub Lo-Boy line with its Model 154, released in 1968. A totally new design of tractor framework was used. This style of frame would be retained by IH until the Lo-Boy was retired in 1979. The IH Farm Equipment Committee Report #249, dated January 11, 1968, and approved on February 23, 1968, released the new International Cub 154 Lo-Boy tractor and its companion 60-inch rotary mower for regular production. The report states, "Trends of buyers and applications indicate a potential for lawn and garden tractors larger than our present 12-horsepower Cadet. This potential exists for a tractor with modern styling and more convenience features at a lower price than our International Cub tractor.

"The present Cub tractor was introduced approximately twenty years ago primarily for agricultural purposes. With changing practices, the demand for a Cub tractor as a farm tractor has declined and the demand for the Industrial Cub has increased. There are certain design characteristics of the Cub that necessitate compromises when used for industrial applications—namely operators' convenience and comfort, visibility, mounting of companion equipment and others."

IH built two prototypes of the 154 Cub Lo-Boy. One was sent to Belleglade, Florida, for experimental field-testing. The other was used for both laboratory and production tooling tests. Laboratory testing was conducted with satisfactory results on both the new engine and the PTO clutches. Production was slated to begin in October of 1968.

This "new" tractor was equipped with a modified IH-built C-60 engine, and a modified final drive unit. The engine was basically the same used in the present Cub tractor, modified to achieve a 15-horsepower output at 2200 rpm by using improved crankshaft bearings, aluminum pistons, improved valves with rotators included on the exhaust valves, and a new manifold. The engine cooling system remained the same. A thermosyphon radiator with an area of 1.4 square feet

IH 154 CUB LO-BOY WITH SICKLE MOWER
The IH 154 Cub Lo-Boy shown here is fitted with a side-mounted sickle bar mower. The mower drives from the tractor's rear PTO under the large cover at the back. This 154 is fitted with an optional hour meter to record the tractor's running time. The meter is the round clock in the center of the tractor photo.
Wisconsin Historical Society

along with a two-row core would allow full engine cooling in 114-degree air. A horizontally mounted muffler with side exhaust outlet was standard equipment. A vertical exhaust was not offered. The higher rpms of the engine help it to develop more horsepower than the previous International Cub Lo-Boy.

The Model 154s were the first Cub Lo-Boys to use a replaceable, dry-type cartridge air filter element. IH used two different types of air filter elements on the Model 154. The Model 185 and Model 184 used an element common to the later Model 154 style.

Because of the higher engine rpms (2200 versus 1800), higher ground speeds could be expected. This would improve the speeds used in mowing operations.

The 154 Cub Lo-Boy differed from its predecessor (Cub Lo-Boy) in that it used a Delco Remy combination starter/generator instead of a Bendix-driven starter. This connected to the engine via a single "V" belt drive between the crankshaft, cooling fan, and starter/generator. The electrical system was of a 12-volt rating. The 12-volt battery was located behind the operator near the right rear fender. Twin sealed-beam headlamps were recessed above the tractor grille. An optional rear combination red-and-white rear lamp was also available. Ignition was accomplished via key-type starting with a safety starting switch on the clutch pedal. The C-60 engine was equipped with an IH-built distributor that had an automatic spark advance.

Because the operator now straddled the transmission (much like on the Cub Cadet) a new steering mechanism needed to be developed. IH used a steering box with a 13:1 ratio. A 15-inch steering wheel along with center point of control of the front axle tie rods achieved a 9.4-foot turning radius without braking. A cast iron, center-pivoted, front axle absorbed shock loads and added rigidity to the tractor.

CUB LO-BOY 154 WITH CUB CADETS
The styling of the 154 Cub Lo-Boy was similar to that of the Cub Cadet line of garden tractors. This family photo with the 154 in the center shows the similarities in appearance. *Kenneth Updike Collection*

The drivetrain of the Model 154 was the same as the current model Cub Lo-Boy. However, the PTO drive is through a pulley on the engine clutch shaft, and a set of matched belts that drive the PTO shaft. This shaft extends under the transmission to a rear clutch to provide a rear PTO. The use of a multiple-disc, dry-type clutch provided a rear PTO speed of 1830 rpm at 2200 engine rpm. The PTO was of an independent type, meaning that it would continue to operate even if the main clutch pedal was depressed. If a tough mowing spot were encountered, the operator could stop the tractor without interrupting engine power to the mower. This is particularly helpful when using a snowblower and inching of the tractor is needed.

IH's Louisville plant encountered a manufacturing quagmire with the 154 that caused about 25 percent of the 154s to suffer driveline failure. The assembly tolerances set forth by engineering were too stringent, causing clutch shaft failures. When the handmade prototypes were built, this defect was not discovered, because production-tooling jigs were not being used.

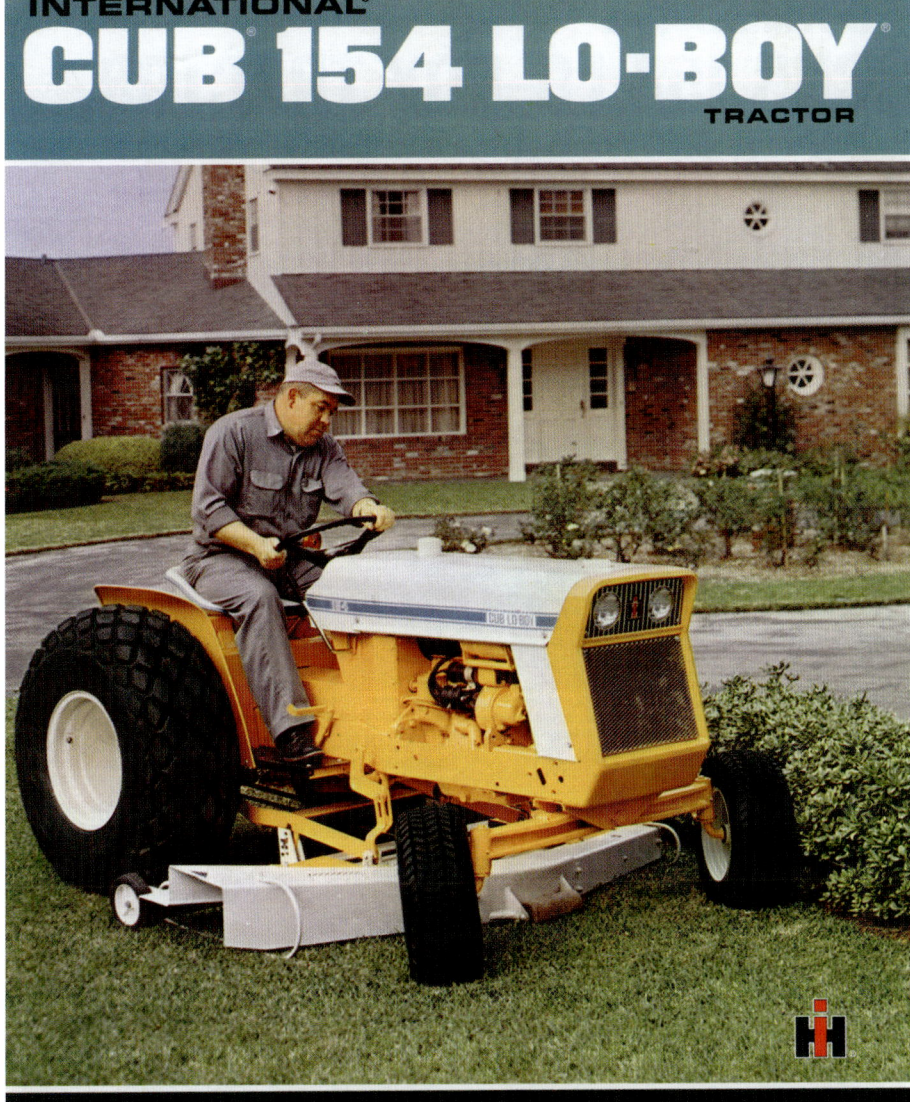

CUB 154 WITH 60-INCH MOWER
This early model 154 Cub Lo-Boy has a 60-inch-wide, three-blade finish mower underneath it. The chromed lever on the right-hand side of the dash operates the hydraulic lift cylinder to raise or lower the mower. *Kenneth Updike Collection*

CUB 154 WITH 60-INCH MOWER
This later model 154 Cub Lo-Boy is identified by its black hood stripe. The gutsy C-60 engine has the power and torque to mow even steep hillsides like this one. *Kenneth Updike Collection*

IH made changes to the assembly processes and tooling, correcting the defective tractors in the inventory and in the field.

The tractor's frame was a major change from previous models in that it was comprised of two full-length, sheet steel-formed channel members. This "framed" Lo-Boy did not use the heavy cast torque tube as the previous Lo-Boys had. By using a straight channel frame, the mounting of allied equipment such as front-end loaders, blades, and brooms would be greatly simplified. The steel disc rear wheels had a fixed-tread design, unlike that used on the previous Cub Lo-Boys. The engine's power was still delivered to the transmission via a driveshaft.

The transmission and final drive were the same as those used on the prior model International Cub Lo-Boy. The 154 came with 4.00-12 ribbed front tires and 8.3-24 rear ag bar tread style tires. Optional 20x8-10 terra front tires and 13.6-16 turf-style rear tires for the tractors commonly set up for yard or lawn work.

The 154 used the same operator's seat as the Cub Cadet 122, 123, 124, and 125. This was solid-mounted with the provision for fore and aft positioning by using selective holes in the seat support. The new seat support had the rear fenders attached to it instead of the rear axle housing for added rigidity.

One of the major improvements the 154 had over its predecessor was the addition of an instrument panel dash. As IH author Guy Fay says, "It's a tractor with *DASH*!!" Here, the engine throttle and choke control levers were located. Both of these were now automotive-styled push-pull cables like that used on the Cub Cadet. The key starting switch and optional horn button were also located here. An oil "Tellite" was also located on the dash to alert the operator of any situations in which low engine oil pressure may be detected. Separate levers to control the rear hitch, hydraulic lift, and independent PTO were also conveniently located within the operator's reach. Because of the restyling changes, the Model 154 looked like it was a modern tractor, because it actually was. This wasn't just an old

INTERNATIONAL Cub 154 Lo-Boy Tractor

CUB 154 SALES LITERATURE

This literature sheet from IH lists the basic specifications of the 154 Cub Lo-Boy. Literature like this was very useful when comparing the 154 Cub Lo-Boy to its competitive tractors. *Kenneth Updike Collection*

New! A work-styled tractor with horsepower, comfort, maneuverability and extra features specifically designed to meet today's needs in lawn, garden and industrial use. The 154 offers a more quiet, water-cooled engine with increased rpm to 2200; independent power-take-off at 1800 rpm; a time-tested and work-proved transmission; cleaner, safer deck; convenient panel-cluster controls; more man comfort—and many other practical features to make it best buy in its class. Team with a 48 or 60-inch mower, front-mounted blade, disc harrow and other tools.

Features and General Specifications

HORSEPOWER:
Horsepower, bare engine (standard sea level conditions: 60°F. temperature and 29.92 in. Hg. barometric pressure dry air)...15.0
Net engine HP per IEMC standard (engine equipped with fan, muffler, air cleaner, generator) (85°F. and 29.00 in. Hg. barometric pressure)................................14.0

(see engine performance curve for torque and HP vs RPM)

Taxable engine hp (A.M.A.)...............................11.0

GROUND SPEEDS, MPH with 8.3-24 R-3 tires
(2200 RPM engine speed, no allowance for slippage.)
Gear: First...2.9
     Second..3.8
     Third..8.6
     Reverse...3.2

TRACTOR DIMENSIONS (with 8.3-24 R-3 and 4.00-12 I-1 tires):
Wheelbase, in...64
Length, overall, in..94
Width, overall, in...52
Height over steering wheel, in..............................49
Ground clearance, minimum (under front axle), in........13
Turning radius, no braking, feet............................9.4
Tread, front and rear, in....................................42
Drawbar height, in..13

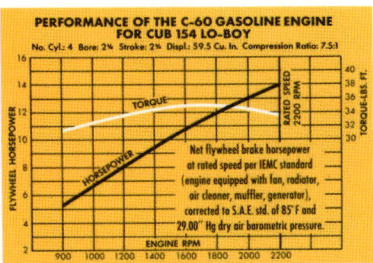

tractor with a new "skin"; IH had made some major design changes to the 154 to make it a new tractor. It is ironic that the 154 was considered an industrial tractor, yet it did not resemble any of the IH industrial tractors of the time. The tractor that was closest in resemblance was the IH 4100 4wd tractor.

A major change with the 154 was its new hydraulic system. Here, an engine-mounted 2-gpm pump rated at 2,000 psi supplied oil to a fixed one-way cylinder to operate the front-lifting rockshaft. A full-flow, replaceable, spin-on type hydraulic oil filter was easy to service, unlike the mesh screen in previous models. A rear-mounted three-point hitch with position control provided a lifting capacity of 450 pounds. A nudging type control valve operated this hitch. This new system— often referred to as Touch Control—replaced the old cast iron cylinder block found on the prior model Cub Lo-Boys. A hand-operated lift providing similar lifting effort to the smaller Cub Cadets was optional.

The 154 offered a choice of either a rigid rear draw bar, or a three-point hitch for tractors that were equipped with hydraulic lift.

Oddly, the 154 Cub Lo-Boy was offered in three different "body styles." The difference in these three types is only the product graphic (decal) packages that were used.

The first style of Model 154 design was used prior to serial number 14535. This tractor had a white hood with a light-blue decal running the length of the hood. The model number 154 was at the rear of the stripe near the instrument cowling. These tractors also have a yellow grille surround and white padded vinyl seat with blue piping trim. They also have a silver-painted mesh grille screen. This was approximately the years of 1968 and the early part of 1969.

The second or intermediate style was used on serial number 14536 to serial number 18708. This was approximately the years of 1969 and 1970. This was painted in the same style as the first, only that the hood stripe was slightly changed.

The final type was used on serial number 18709 and above. These styled Cubs were built from 1970 until the Model 154 was replaced with the Model 185 in 1974. Here, the 154 used a black decal stripe with a blue pinstripe border. A vinyl padded black seat and vertically ribbed front grille screen are also characteristic of this model variation.

On all of the 154 Lo-Boy tractors, the basic styling of the hood and grille remained the same. Only the accentual striping was changed to match the "current" production Cub Cadet striping scheme. This gave the 154 a family-appearance look with the Cub Cadets.

IH had planned to not only test, but also market (at a later date) a hydrostatic drive transmission in the 154. IH noted that a complete hydrostatic transmission of the size and cost needed for the 154 was currently not available in the industry. Hydro vendors would, however, approve the use of current units when combined with the usage of supplemental gearing. The 154 hydro would be a "component" tractor. Here, the tractor manufacturer (IH) would use parts or components from outside vendors in combination with its own manufactured parts to build the final product. The exact specifications of this tractor elude this author at the time of this being printed. However, one could venture to guess that if IH engineers wanted to take this tractor "one step further" they easily could have. The idea this author is alluding to is a Cub Lo-Boy with a hydro-mechanical transmission. Here, the three-speed transmission would be coupled to a hydrostatic drive unit. This would produce infinitely variable speeds in three ranges. The reverse gear would be removed from the mechanical transmission, as the hydro would control the machine direction. For some reason, the most likely being cost of production, IH never did pursue the Lo-Boy Hydro tractor version.

Another odd version that IH planned but never made was the Model 154 High-Clearance tractor. While the specifications of this machine have yet to be found, this author would venture to estimate that the 154 Hi-Clear *could have* replaced the International Cub. The 154's hood and grille styling would have been retained; however, the "rail frame" of the 154 may have been replaced by the cast torque tube that the Cub used to connect the engine to the transmission. IH studied offering this tractor in 1967, but it was never built. No evidence exists that IH actually built a 154 Hi-Clear prototype tractor, either.

IH built 29,171 copies of the 154 Cub Lo-Boy during its production run from 1968 to 1974. None of the "numbered" series of Cub Lo-Boys (154, 185, 184) were ever tested at the Nebraska Tractor Testing facility in Lincoln, Nebraska. The original list price for a 154 Cub Lo-Boy was $1,788 in 1968. These were sold exclusively through the 3,600 IH dealers located nationwide in the United States and several foreign branches.

IH's competitors were offering "garden tractors" in the 14-horsepower range, when compared to the Cub Lo-Boy Model 154. John Deere was selling the Model 140, which was a 14-horsepower garden tractor that had been upgraded from a 12-horsepower model in 1969. Bolens offered its Model 1455, which was basically a 14-horsepower upgrade from their 12-horsepower Model 1250. J. I. Case sold its Model 444 hydraulic drive tractor. This offered no basic design changes from its 12-horsepower Model 442. Wheel Horse sold its Model GT 14 against the Model 154 Cub Lo-Boy. This was smaller than the 154 and offered a 48-inch mower deck. None of the other manufacturers had a model that compared directly to the 154. IH once again had discovered a niche market.

Model 185 Cub Lo-Boy

The 185 Cub Lo-Boy would be the last yellow/white-painted Lo-Boy tractor that IH would sell. It was styled similarly to the Model 154, in that the hood and grille shell retained the same basic shape as the 154. The grille screen and hood striping was changed, along with the grille shell now being painted white to match the hood. The front grille and headlight panel were painted black, and a single blue pinstripe paralleled the larger black stripe on the hood's side indicating this was the Cub Lo-Boy 185.

The 185 was also IH's first high-horsepower Cub Lo-Boy. The IH-built 3/4-inch updraft carburetor, found on the previously built 154, and all of the older Cub Lo-Boys and Cubs, was replaced by a new Zenith-brand carburetor featuring larger metering jets and a larger air-filter-to-carburetor rubber air intake hose. Domed cast aluminum pistons inside the engine helped burn the added fuel more completely. With more air and fuel in the engine, more horsepower can be created. The 185 also had the governed rpm limits of the engine adjusted upward to build

185 CUB LO-BOY
IH's 185 Cub Lo-Boy was a sales success! The 185 replaced the prior Model 154 Cub Lo-Boy tractor in 1974. IH made over 6,300 units of the 185 Cub Lo-Boy. *Lee Klancher*

185 CUB LO-BOY TRACTOR ENGINE

The 185 Cub Lo-Boy is fitted with a Zenith-brand carburetor shown here as the large silver object in the photo. The 185 also has a horizontal exhaust that exits toward the front. *Lee Klancher*

Cub Lo-Boy
185

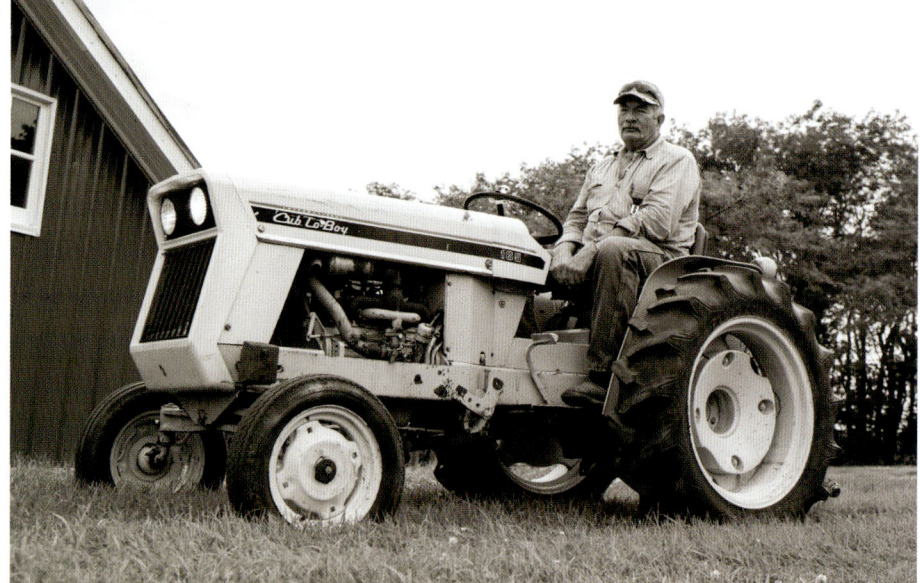

REAR THREE-POINT HITCH ON 185

IH fitted the Cub 185 Lo-Boy with an optional Category 1, rear three-point hitch. The hitch was operated by moving the hydraulic lift lever, and it raised and lowered with the center rockshaft. *Kenneth Updike Collection*

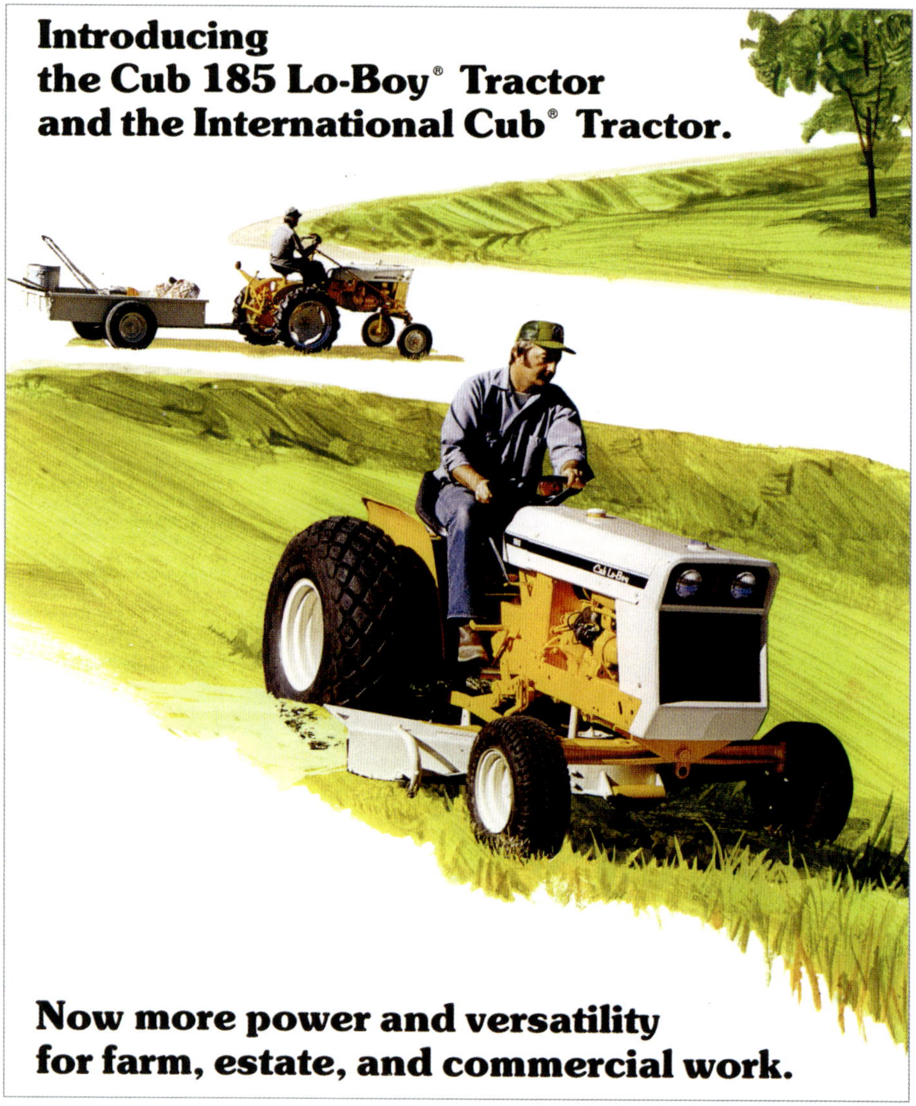

Introducing
the Cub 185 Lo-Boy® Tractor
and the International Cub® Tractor.
Cub Lo-Boy
Now more power and versatility
for farm, estate, and commercial work.

CUB LO-BOY TRACTOR
The low ground profile of the Cub Lo-Boy 185 meant the tractor was very stable to operate. The tractor has three forward gears and one reverse gear in the transmission, and features independent PTO and hydraulic lift. *Lee Klancher*

more horsepower, too. The low idle speed was set at 600 rpms versus the Model 154's setting of 475 rpms. The high idle (2510 rpms) and the rated load speed (2300 rpms) were both boosted over the 154's settings of 2420 rpms at high idle and 2200 rpms at rated load speed. By turning the engine faster, more horsepower can be created. The downside to this is that as engine rpms increase, the piston travel speed in feet per minute increases, too. This increased piston speed translates into increased engine wear. In short, higher engine rpms mean faster engine wear.

The Model 185 rear brakes were equipped with either a single pedal or dual rear brake pedals. The dual brake pedal kit was available as an optional extra cost item.

The 185 was painted IH 483 Federal Yellow on its chassis and rear fenders. The hood, grille housing, and wheel rims were IH 935 White. The upper and lower grille panels, footrest boards, and steering column were painted IH Black. IH located the serial number tag (part number 166058C1) on top of the LH frame rail near the front axle. This made identifying the tractor for service parts components much easier.

Nearly all of the implements offered by IH with the Model 154 were carried over into the 185's production, too.

The International Model 185 Cub Lo-Boy had 6,346 tractors built during its relatively short production span from 1974 to 1976. The average retail price of a 185 Cub Lo-Boy in its last year of production (1976) was $3,820.

1975–1979 Cub

In 1975, IH made its last major model change to the Cub tractor. This series was called the New International Cub Tractor (with increased horsepower). These Cubs started production very late in 1975 at serial number 248125 and continued until 1979 with the final built Cub serial number 253685. This meant average annual production of the model was 1,853 tractors. Compare that to the original Cub, which had an average annual production of 20,388 units in its first ten years. The last Cub experienced a drastic decline! All told, only 5,561 units of the the new International Cub were built, making the model one of the lowest production Cub tractor variations made. The styling of these Cubs remained unchanged from the previous International Cub except that a long, black stripe with an accenting blue pinstripe were used on these machines. The basic machine was still bathed in IH 483 Federal Yellow paint with the hood, grille shell, lights, and wheels being painted IH 935 white. The operator's seat was painted black. IH sold these Cubs with English, Spanish, French, and German caution/warning decals on them to meet the standards of the various countries they were exported to. The increased horsepower Cub was listed as producing 15 horsepower in IH sales literature of the time.

IH continued to offer its Touch Control hydraulic system and its patented rear one-point Fast Hitch until the end of Cub production. IH was experiencing a high number of warranty claims on the Cub's hydraulic system due to poor castings.

The increased-horsepower International Cub used the same basic engine specifications as the Model 154 Cub Lo-Boy tractor.

In 1978, IH offered three compact tractors to the market. The 15-horsepower International Cub, the 18-horsepower 184 Lo-Boy, and the 28-horsepower model 284 compact tractors. The 200-series compacts that IH would introduce in the early 1980s replaced all three of these models. With the financial problems IH was having, coupled with new proposed laws affecting exhaust emissions on tractors, it was clear that the Cub was near the end. The last published list price of the International Cub was in 1979, and it listed for $5,350. Some Cubs today can still command upwards of this price nearly thirty years later.

IH CUB TRACTOR WITH MOWER

The IH Cub, when fitted with a 60-inch-cut under-mounted mower, is ideal for cutting large areas of turf, such as parks and the like. The twin-gauge wheels on the mower reduce scalping. The mower is belt driven from the tractor's rear PTO.
Wisconsin Historical Society

IH CUB TRACTOR
The use of diamond-tread rear tires on this Cub tractor minimizes turf damage. The three-blade 60-inch-cut mower is lifted with the tractor's hydraulic system by the silver metal chain. *Lee Klancher*

CUB WITH MOWER
Mowing grass at a commercial business is a fast task for the IH Cub when it has a 60-inch-wide, three-blade mower underneath it. The mower is belt-driven from the tractor's rear PTO. *Kenneth Updike Collection*

CUB TOUCH CONTROL LIFT

The use of lifting implements with the Cub is near effortless when the Touch Control lift is used. The simple one-lever control has a sliding depth stop that is adjustable. *Kenneth Updike Collection*

CUB PLOWING SNOW

Moving snow is easy when a Cub tractor and front blade are involved. The large spring above the blade presses the blade to the ground. Another spring allows it to trip and spring back into working position if an object is hit. *Kenneth Updike Collection*

CUB AT HINSDALE
This photo was taken at IH's Hinsdale Engineering Center, located in Hinsdale, Illinois. The tractor is parked on a giant rotating turntable to allow photos to be taken from nearly any angle. *Kenneth Updike Collection*

OPERATOR'S PLATFORM
The view of the operator's station on a late-model Cub shows the amp gauge and rotary light switch. Below these two is the key starting switch. The black button under the key switch is the fuse to protect the electrical system. The engine choke cable is the black knob on the left side. *Kenneth Updike Collection*

Model 184 Cub Lo-Boy

The Model 184 Cub Lo-Boy brought the use of IH Red paint back to the Cub tractor line. The 184 was styled after its "big brother" lineup of farm tractors, the IH 86 series.

The basic tractor was bathed in IH 2150 Red, with IH 935 White wheel rims and side hood panels. A black decal stripe ran the length of the hood side and was similar in format to the larger IH 86 series farm tractors in design. The grille is styled as a scaled-down version of the IH 686 tractor grille.

The characteristics of the C-60 engine on the 184 were similar to that on the prior Model 185. Cub engines with serial number 312390 and above were equipped with a Zenith updraft carburetor. The 184's cooling system was identical to the previous models except for the radiator cap.

A newer Zenith carburetor that contained a revised needle valve and seat along with a larger orifice was added February 28, 1977, at serial number 43801. This change was made to overcome field complaints of the 184 having vapor lock and stalling out found during prototype testing.

The C-60 engine in the 184, featured heavy-duty construction, designed to be long lasting. A cast iron crankcase and cylinder head, along with a forged-steel crankshaft that has induction-hardened bearing journals (for long life) were standard issue on the 184. The engine main and connecting rod bearings featured an exclusive tri-metal design that allowed them to be serviced in the field with replaceable insert-style bearings. Aluminum-domed top pistons with two compression and one oil ring were the same as those on the previous Model 185. Using alloy

184 TRACTOR WITH BLADE
The IH 184 Lo-Boy tractor is ready for any necessary task. In the winter, snow removal is a quick job when the 184 is fitted with a front-mounted blade. *Kenneth Updike Collection*

IH 184 LO-BOY TRACTOR WITH TRAILER

IH's 184 Lo-Boy tractor is helpful with many chores around the farm or ranch. Building or repairing a fence is much easier when a 184 and IH trailer are used to move materials. *Lee Klancher*

CUB 184 FRONT AXLE

The massive cast iron front axle of the 184 Cub Lo-Boy is built for long life and durability. The I beam-style casting has zerk fittings on the axle spindles that allow them to be lubricated with grease. *Kenneth Updike Collection*

CUB 184

A vintage Cub advertisement. *Kenneth Updike Collection*

steel intake valves and Silchrome steel exhaust valves (with rotators on exhaust only) that were resistant to valve scorching would allow the C-60 engine to still operate properly using unleaded fuel that was fast replacing regular gasoline. The lead in regular gasoline was used to lubricate the engine's valves.

To keep the 184s running cool, the fan hub and blade assemblies underwent a major revision in September of 1979. Here, the old-style fan/hub combination would now be offered as individual parts: a fan hub, and a fan blade assembly. The design of the fan hub was changed from an oil bath shaft, bushing, and sheet-metal pulley, to a new cast iron pulley and replaceable sealed ball bearing. This new-style hub fit not only the 184, but also was retrofitted to the Cub, Farmall A, B, BN, A1, AV, AV-1, Super A, and C tractors.

The earliest production 184s used a combination starter/generator like the 185 model had. A new and improved starter/generator first saw installation on November 8, 1977. This new starter/generator had a lighter-weight frame, aluminum conductor field

CUB 184 AT WORK
The front-wheel hub of the 184 is cast iron for extra strength and has press-in wheel studs to attach the wheel rim to it. *Kenneth Updike Collection*

CUB 184 FRONT GRILLE
The front grille of the 184 is molded in fiberglass. The styling of the 184 actually matched the larger IH tractors of that time. The 154 and 185 were not styled to look like the larger IH tractors. *Kenneth Updike Collection*

coils, and an improved solenoid switch. At serial number 46112, IH began using a direct drive starter, which engaged a flywheel-mounted ring gear. This directly replaced the older starter/generator design with a separate cranking motor and alternator. A Delco brand 42-ampere alternator with an internal solid-state regulator was added to handle the tractor's charging duties. The 184s electrical system was of a 12-volt design.

One of the littlest parts that a person would expect to cause problems did. The ignition key was prone to breakage. IH used the same ignition key in most of its tractors for parts commonality. The IH 86 series farm tractors had the ignition key "handily" positioned at operator knee height. When entering or exiting (or just turning your body), your knee would hit the key. A similar problem arose in the 185 and 184 Lo-Boys. IH fixed this by replacing the "thin" key (#382458R2) with its new "thick" key (#131313C1) at serial number 47045 on the 1978 production year Model 184. The 131313C1 key has a stronger cross-sectional area and is still used as the standard service part today.

Basic frame design of the 184 remained unchanged from the previous Model 185 design until December 13, 1977. IH modified the frame to accommodate the new flywheel-mounted starter. The flywheel shield was modified at this same date for the same purpose.

The transmission used on the 184 was identical to the previous Model 185 except that the main clutch driveshaft now used a universal joint at the transmission main shaft instead of a flexible disc drive coupler. The clutch release bearing yoke was also improved over the 185 with a stronger Y-shaped design used.

The 184 was equipped with a single brake pedal, but a two-pedal brake was available as a parts accessory. The 184 still used band-style brakes, which contacted cast metal drums to stop the tractor.

A creeper gear reduction drive gearbox was an optional attachment on the 184, like it was on the 154 and 185. This was recommended with the use of a front-mounted snowthrower or rear-mounted tiller. This gearbox had to be manually engaged via a shifting lever with the tractor's clutch pedal depressed to prevent gear "clash." The creeper gear drive box that was used on the Model 184 differed from the Model 154 and Model 185 in that it used a Woodruff key to secure the U-joint to the creeper drive housing. Since the 184 transmission shaft drives with a Woodruff key and the creeper drive output is pin drive, the transmission driveshaft must be replaced also when installing this attachment.

The Model 184 used a rear-mounted electric clutch to activate the tractor's PTO. This clutch was similar to that used in the current production Cub Cadet garden tractors of the time. By using this, the operator could truly have an independent PTO. The main drive clutch no longer controlled all of the tractor's PTO functions, too. IH engineers noted that the possibility existed that the front and rear PTO shaft support bearings could fail at an early age. To correct

CUB 184 GRILLE EMBLEM
The plastic IH emblem used on the grille of the 184 was attached with two screws from behind the grille. This gave the front end a clean, bold look.
Kenneth Updike Collection

CUB 184 GAS CAP
The Cub 184 had an available snow plow. With a set of chains, the little tractor was quite effective at this task. *Kenneth Updike Collection*

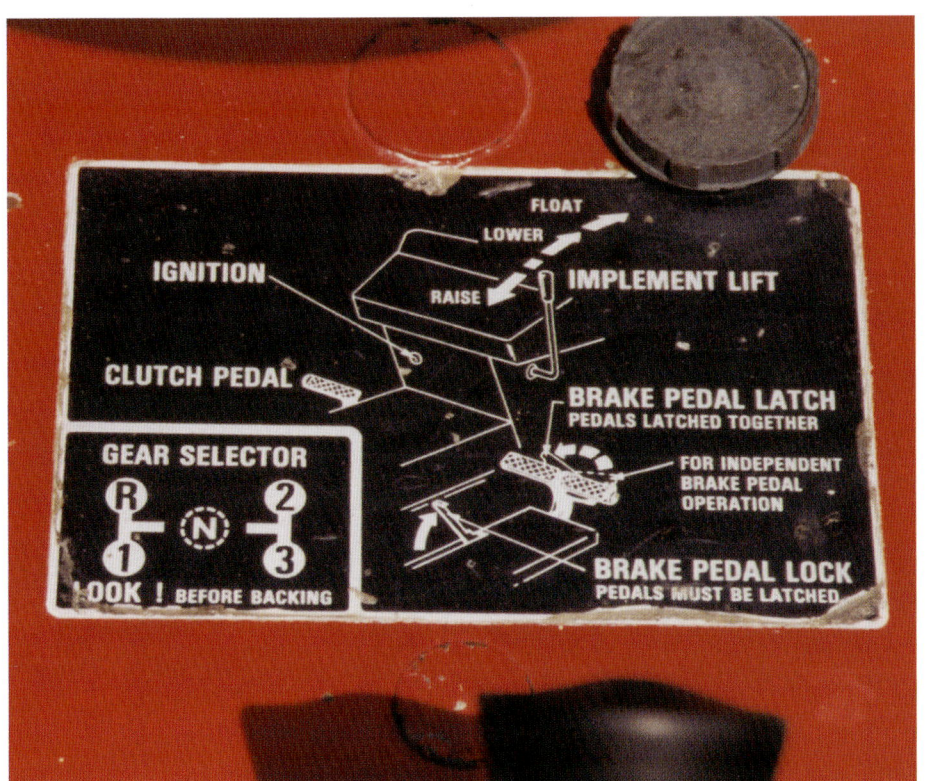

CUB 184 DECAL
The instructional decal of the 184 is located on the tunnel cover over the driveshaft. This decal illustrates the dual brake pedal version. A single brake pedal was also offered. *Kenneth Updike Collection*

CUB 184 REAR VIEW
This rear view of the 184 shows the Category 1 rear three-point hitch and the lifting cylinder that operates it. Under the white guard is the rear independent PTO pulley that powers the mower.
Kenneth Updike Collection

this, the PTO shaft diameter was increased, as was the related bearings at serial number 45004. The use of larger bearings also made to provide better bearing contact to the shaft. The PTO operated at a full engine load speed of 1880 rpm.

The competition offered garden tractors to compete against the 184 in the marketplace. IH was quick to capitalize on this and used the 184's advantages in advertising the better features that could only be found on the 184. To compete with the 184, John Deere offered its model 400. This was more of a super garden tractor in size than a compact tractor. The 400 more closely matched IH's soon-to-be-released 982, than it did the IH 184. The Deere 400 was powered by a twin-cylinder, air-cooled Kohler engine rated at 19.9 horsepower. Bolens offered its model HT23 that also was Kohler twin-cylinder powered like the Deere, but had an engine rating of a whopping 23 horsepower. Simplicity built both its model 9020 and the Allis-Chalmers 720 tractor. These tractors were basically identical to each other except for the paint scheme, decals, and body styling (hoods). These two tractors were both powered by twin-cylinder, air-cooled, Onan engines that were rated at 19.35 horsepower. All of the competitive models mentioned were gasoline powered.

All of the competitive models were equipped with a hydrostatic transmission, a hydraulic implement lift, and a 12-volt electrical system. The Allis-Chalmers 720 and the Simplicity 9020 were the most comparable to the 184. They both were the same approximate physical size as the 184. The Bolens HT23 and the 400 Deere were more scaled to the 982 super garden tractor that IH offered.

Even though the 184 was underpowered when compared to the competition, it overpowered them with its features that they lacked. With an 8-gallon fuel tank, the 184 easily outran the Bolens HT23,

184 OPERATOR VIEW
The view from the operator's seat shows the gearshift (lever in center), clutch pedal on the left, and dual brake pedals on the right. The steering column is offset as it is mounted to the frame rail underneath.
Kenneth Updike Collection

which offered only a 5-gallon supply. The 184 also offered an optional Category 1 rear three-point hitch; all of its competitors could only offer a smaller Category "0" hitch. The competition did outrun the 184 in that they all were equipped with only a hydrostatic drive transmission versus the 184's 3-speed gearbox. As previously noted, IH at one time investigated building a Hydro Cub Lo-Boy, but never marketed one.

The marketing department at IH intended to sell the 184 primarily to commercial users such as landscapers and turf/groundskeepers. By stressing the advantages of IH's smooth running, water-cooled, efficient engine along with IH's famous durability, it would be a natural to choose the 184. A key selling point was that IH's water-cooled engine lasts at least twice as long as those offered by the competition's air-cooled counterparts. Why would you want to sacrifice service reliability?

IH built 4,228 units of the International 184 Cub Lo-Boy in its 1976 to 1979 production. When compared to the 6,346 units of the prior Model 185 (built from 1974 to 1976), the 184 production was not record setting. This reduced production makes the 184 a highly sought-after modern-day collectible tractor. It is not uncommon to find 184 tractors selling for a higher price now than when they were new. It is ironic in that the styling of the 184 Cub Lo-Boy was similar to that of its larger model farm tractors, the IH 86 series, just as the soon-to-be-released 82 series Cub Cadets would emulate the IH 50 series tractor styling. When the little tractors finally returned to looking like their big brothers, their production ended.

Foreign-Made Cubs

The market for the Cub tractor was outstripping the production capacity of the Louisville, Kentucky, factory that produced them. IH's biggest market for the Cub (after the United States) was Europe. Postwar Europe needed to mechanize their farming methods quickly and on a large scale. IH started exporting Cub tractor components to France in 1949 and continued to do so through 1954. The very first Cubs built in France used mostly components from the United States except the electrical system. Here, French-sourced Ducellier and Paris Rhone brand electrical components were used. The Cub was a 6-volt electrical system. The French-built Cub also has a different carburetor than its US-built version does. In France, a larger Solex-brand carburetor was used.

The original made-in-France Cub tractor was built from 1954 to 1957. These tractors were marked as the FFCUB on the model tag, meaning French Farmall Cub. The French-built Cubs have a starting serial number of 700000.

The French Cub tractor has an F-235 tractor series seat suspension. The boxcar spring/pogo stick suspension used on the US-built models was *not* used on the French-built Cubs. Horizontal exhaust was a standard feature on these tractors too! The IH Fast Hitch rear implement attaching system that was used in the United States was never offered in France.

In 1958, the French Cub was improved. A new version that had a restyled front grille was introduced. This was called the new Super Cub. These tractors were tagged as FSCUB, meaning French Super Cub tractor. Only the French-built Cub tractors had this

FRENCH SUPER CUB TRACTOR
The famed French-built IH Super Cub tractor, shown here fitted with a mounted cultivator. Notice the "McCormick-International/Farmall Cub/Super" lettering on the hood to distinguish this model from other tractors. *Jean Cointe Collection*

FRENCH CUB TRACTOR
IH collector Mike Schmulach imported this French-built Cub to North America. The French-built Cub was never sold in North America. *Lee Klancher*

FRENCH CUB SERIAL PLATE
Close-up view of the French Cub tractor's serial number plate. The plate is affixed to the front bolster with four rivets like the American Cub. The plate lists the tractor serial number (700812) and states that the Cub was built in France. *Lee Klancher*

FRENCH CUB ENGINE DETAIL

A side view of the French-built Cub showing the stainless steel emblems and the C-60 engine. The French-built Cubs have a different electrical system than that of the American-made Cubs.
Lee Klancher

model name (Super). This tractor series started at serial number 720000. The engine horsepower was increased on the Super Cub by increasing the governed engine speed to 2000 rpm. The Super Cub also has a foot throttle, and it was offered with hydraulic lift. In 1961, the grille color was changed to white from red. The US-built Cub tractors oddly never had a "Super" version built. Cub tractor production in France ended in 1964. The Super Cub oddly has the same US-styled front grille as the 1955–1958 model US-built Cubs have.

A Cub Narrow model variation was also built. This tractor has a much narrower wide front axle. Instead of having a cast tube housing to offset the right-hand final drive, the final drive is bolted directly to the rear differential housing as was the left-hand side.

This gave the Cub Narrow a much, much narrower overall tread width. For cropping on slopes and such, the Cub Narrow would not be a good choice of tractor to use. However, when used in narrow crop applications like vineyards and such, the Cub Narrow was the tractor of choice to use!

IH did offer Cub-specific-sized implements/attachments for the Cub, too! An F189 plow and F22 mower were built from an American design but were produced in France at IH factories in Montataire and then later in Croix.

Cub production in France continued until 1964 when production stopped. The French-built Cubs were the only Cub tractors made by IH that were not produced at the Louisville plant.

FRENCH CUB LICENSE PLATE

Close-up view of the tractor license plate of a French-built Cub. In many European countries, tractors are registered and licensed like cars and trucks. This license plate is fitted to a special bracket that also has a small red taillight fitted to it. *Lee Klancher*

FRENCH CUB TRACTOR

The French-built Cub tractors are fitted with a horizontal exhaust system. The French-built Cub was painted IH Red. *Lee Klancher*

CHAPTER 3

Getting Technical with Your Cub

The Cub's technology (such as it is) is broken down in the section below. Read on to learn about the systems that make your Cub a hard-working addition to your household. The Farmall Cub was a sleek, modern compact tractor that came out in the late 1940s well before compact tractors were in style. It offered its owner modern conveniences like electric starting and easy to use hydraulics at an affordable price. IH wanted the Cub to replace the horse or mule on every farm that had one still working on it yet. The modern styling and ease of maintenance made the Cub a sales success!

Engine

The Cub was powered by an IH four-cylinder, thermosyphon-cooled, gasoline engine. This engine was of a "bore in block" design, which meant that the engine does not have replaceable sleeves or wear liners. A 2 5/8-inch bore by 2 3/4-inch stroke yielded a 59.5-cubic-inch displacement. The engine was called the C-60. The "C" represented the word "carburetor" and "60" was the cubic-inch displacement of the engine. IH used this engine identifying method on all its engines. The C-60 engine was of an "L" head–type design. This meant that the valves were not located above the pistons but in a "pocket" area beside the combustion chamber. Only the spark plug was located above the piston.

Running at a rated speed of 1600 rpm, the C-60 produced a whopping 9.25 horsepower on the belt and 8 horsepower at the drawbar. This was no Model H or Model M tractor—it was a "li'l one."

IH built the C-60 engine "in house" on a separate assembly line at the Louisville plant, using the latest design technology of the time. Thousands of hours were spent on the engineering, design, research, and testing of this engine. The C-60 was state of the art when it was introduced. Its basic design remained virtually unchanged in its thirty-plus-year production life span. The crankshaft and pistons were a new engineering design. The original Cub offered cast iron flat-top pistons. In the late 1960s, IH changed these to cast aluminum domed-top pistons. These pistons helped the C-60 engine jump from 10 horsepower to 15 horsepower and eventually to 18 horsepower. When the C-60 engine was outfitted with aluminum domed pistons, the engine compression ratio soared from 6.5:1 to 7.5:1.

IH engineers also made modifications to the camshaft for added durability at higher-rated engine speeds. The camshaft is a single-piece drop-forged shaft with three bearing journals machined into bores in the engine's crankcase. The valve tappets were also improved with the new cam. The Cub crankshaft was drop-forged, high-carbon steel, and were statically and dynamically balanced. The bearing journals were induction hardened. The connecting rods were made of heat-treated metal of an I beam-type design for maximum strength.

IH did offer a high-altitude cylinder head as optional equipment, along with exhaust valve rotators for the Cub engine.

To control the engine's speed, IH engineers outfitted the C-60 engine with a simple fly ball variable speed-type governor. The governor depends on centrifugal force made by weights that rotate on a shaft. These weights are counteracted by a variable tensioned spring. As the weights move, they also control the movement of a connecting linkage that controls the tractor's throttle opening in the carburetor. When the tractor is started and the operator advances the engine speed lever forward, the governor weights move outward by centrifugal force until the governor spring counteracts their movement. A weak spring

LEFT-HAND C-60 ENGINE DETAIL
Left-hand view of the C-60 IH-built engine that powers the Cub. The horizontal exhaust exits to the rear of the engine and goes downward to the muffler. The round breather behind the carburetor serves as the engine oil fill. *Lee Klancher*

RIGHT-HAND C-60 ENGINE DETAIL
Right-hand view of the C-60 IH-built engine that powers the Cub. The black box in the photo is the voltage regulator. The engine coil and distributor is located to the left of the voltage regulator, behind the steering rod. *Lee Klancher*

(very common with use and age in any tractor) will produce less resistance and the engine's performance will be noticeably slower, sometimes nonresponsive (broken springs are characteristic of this). Once the governor spring "stabilizes" the air/fuel mixture of the carburetor, it will be sufficiently regulated by the throttle shaft valve to maintain the engine speed. The operator controls the engine speed by moving a single lever, which increases or decreases tension on the spring, not by a direct connection to the carburetor throttle valve. By increasing the spring tension, the weights move inward, which in turn opens the throttle shaft valve, further increasing engine speed until the spring equalizes the force of the weights.

To slow an engine down, this mechanical process is reversed. The linkage between the governor and the carburetor is adjustable to achieve full power at rated engine rpm. The engine governor drive gear also acts as the ignition unit drive. The governor drive gears are marked for proper mesh with their mating gears at top dead center of the number one cylinder on its compression stroke. If the governor is removed, be sure the engine is timed to this position first.

Engine Oil System

The Cub's C-60 engine was unique in that it had an oil filter housing cast into the engine block. Only the filter lid was removed to service the cartridge-type oil filter. This cartridge-type filter could trap dirt particles as small as .000039 of an inch. The Cub was first equipped with a Purolator brand "umbrella-" style filter. Later in the Cub's production life (1960s), this umbrella style would be changed to a true canister-style filter. The canister-style filters offered a larger filtering surface area to clean the engine oil more efficiently. A spin-on style filter was never offered in the thirty-plus-year production of the Cub. The engine oil pressure gauge screwed directly into the filter housing. The engine was filled with fresh oil via a tube located on the engine's left side. When the breather cap was removed from the tube, the oil level bayonet, or "dipstick," that was attached to it, showed the current level of oil in the engine. It is important to check the engine oil level *every* time before the tractor is started. If the level is low, add sufficient oil to bring the level into the "safe" range indicated on the stick.

Pressure lubrication of vital engine components was accomplished via the use of an engine-driven oil pump located in the rear of the engine. A fixed oil pickup screen in the oil pan sump fed the pump. After going through the oil pump, the oil is forced through drilled passages to the crankshaft bearings, the connecting rod bearings, and the camshaft bearings. All of the critical engine-bearing surfaces are protected continually and positively with a light film of friction-reducing oil.

The Cub had an engine oil capacity of three quarts. IH recommended that the engine oil and filter element be replaced every 120 working hours of operation. This author recommends that the engine oil be changed every 100 hours of operation and that only straight 30-weight (10-weight in colder climates or winter use) Lo-Ash oil (made by IH, now CNH) be used. The reason to use Lo-Ash oil from IH is that this oil has a lower sulfated ash content than other "off-the-shelf" oils do. This lower ash content can reduce valve deposits and torching. This results in a cleaner-burning engine. When the original Cub owners' book was printed, it stated to use a 30-weight *non-detergent* type of lubricating oil in the engine. While this may have been state-of-the-art lubricant in the 1940s, oils today have been vastly improved and every engine should be treated to them. Many times Cubs are bought used or secondhand and the oil type is unknown. An easy way to see if your Cub has detergent or non-detergent oil inside is to remove the engine oil pan or the side cover accessing the valve tappets. If large accumulations of "sludge" are present, the tractor probably has non-detergent inside. If the areas are clean, detergent oil has been used. Because detergent oil holds the dirt particles in suspension until they are filtered out, this is logically the better choice. The C-60 engine used on the Cub series tractors *has* a replaceable oil filter element. A detergent-type engine oil should be used to carry the dirt and contaminants to the filter so they can be cleaned from the engine oil. Changing a non-detergent engine to detergent oil can require multiple filter and oil flushing. The detergent oils will loosen the deposits inside the engine and move them to the filter eventually.

As engine and oil technology changed, so did IH's recommendation for the engine oil. A special Low-Ash oil was developed by IH that has a lower sulfated ash content than other engine oils. The engine oil for a diesel engine and its gas counterpart have significantly different additives or chemical compositions. Typically, diesel engine oils require a higher sulfated ash content; this is vital for lubrication. When higher ash oils are used in a gas engine, serious damage can occur, such as valve torching and valve stem deposits. IH engineering discovered this and they formulated a low-sulfated-ash-content oil called (what else?) Lo-Ash engine oil. Since Lo-Ash oil has a lower sulfated ash content, longer engine life with fewer issues are part of the benefits derived from using it. IH recommends the use of Low-Ash engine oil in the Cub and Cub Lo-Boy for longer engine life.

CUB OIL FILTER
The round canister located just below the steering rod that has the yellow decal affixed to it is the engine oil filter. This is a replaceable cartridge-type filter that can be accessed by removing the cover lid on the filter housing. *Wisconsin Historical Society*

Fuel

One feature that was commonly found on other tractors made before and after World War II was a choice of fuels used to power the tractor. IH offered tractors that would burn diesel, gasoline, or kerosene. The Farmall Cub was only offered with a gasoline engine. The Cub could run on kerosene (distillate), but it lacked a carburetor with an adjustable main jet, a separate gasoline starting tank, and the radiator shutter attachment commonly used on tractors that ran on kerosene. A clear glass sediment bowl under the fuel tank not only filtered out water and other impurities, but with a simple glance the operator could check to see if the system was dirty. The engine's fuel system used a gravity-fed, IH-built, 3/4-inch, updraft carburetor. A secondary screen at the carburetor's fuel inlet trapped any dirt that the sediment filter may have missed.

FUEL LINE CONNECTION AT CARBURETOR
The steel fuel line from the gas tank to the carburetor being connected to the carb. The fuel is gravity fed from the tank to the carb, meaning no fuel pump is needed. *Lee Klancher*

Carburetor

An IH-built carburetor was used on the 1947–1974 Cub and Cub Lo-Boys. You can identify this carburetor by the IH emblem cast into the carburetor bowl. The 1975 and later Cub and Cub Lo-Boy use a carburetor manufactured by Zenith. This carburetor has "Zenith" cast into the body. This makes identifying the two carbs very easy. The French-built Cub and Cub Lo-Boy used different carburetors. If you find any other brand of carburetor (e.g., Marvel-Schebler or Holley) fitted to the tractor, it is not a factory-built setup.

IH C-60 CARBURETOR
The IH-built updraft carburetor for the C-60 engine was of a single-barrel design. IH built these carburetors themselves, as they did many of the components used on the Cub and Cub Lo-Boy tractor. *Lee Klancher*

IH C-60 ENGINE CARBURETOR
At the factory, a Cub's IH-built carburetor was painted red. *Lee Klancher*

Ignition

IH designed a magneto ignition specifically for the Cub. The heart of this system was the J4 magneto, which was standard equipment on Cubs. IH offered its own distributor ignition as optional equipment beginning in mid-1950. The distributor was considered standard equipment when the "battery ignition" package was ordered. All of the distributors IH used on the Cub were similar to those used on larger farm tractors. The main difference was the distributor driveshaft had a different timing advance made into the shaft and different advance springs. These distributors have an automatic spark-advance feature. This package offered battery-powered starting, two headlamps, one rear work lamp, and distributor ignition. Starting was accomplished via a Bendix drive 6-volt starter made by the Delco Remy Company. This starter engaged a metal-toothed ring gear on the Cub engine flywheel. The starter was located on the right-hand side of the tractor directly below the engine oil filter housing. Electric starting and lighting became a standard feature in 1958.

6-Volt versus 12-Volt Systems

In the mid-1950s, IH switched all their engines from a 6-volt electrical system to a 12-volt electrical system. The use of 12-volt electrical systems was becoming more common in ag, industrial, and

IH DISTRIBUTOR IGNITION
The IH-built distributor on the Cub and Cub Lo-Boy was used on tractors with factory-fitted electric start. The black distributor cap and the engine oil filter were left unpainted at the factory. *Lee Klancher*

automotive applications. IH followed suit with the Cub. The IH parts books for the Cub and Cub Lo-Boy cover the various wiring harnesses used on the tractors. They list the serial number breaks when the changes occurred on the tractor changing from 6 to 12 volts.

The Cub was right in the middle of that, so you have to check to see which system is on your machine. Bear in mind, many 6-volt machines were later upgraded to 12-volt systems, so don't assume your 1947 model still has the 6-volt system intact.

The easiest way to tell what electrical system your Cub has is to look at the battery. If the battery has three or four filler caps, the electrical system is 6-volt. Note the four-filler cap battery is an 8-volt battery, which would work with a 6-volt system. That battery is now fairly rare.

If the battery has six filler caps, it is a 12-volt. Another quick method to identify what the tractor may be (6- or 12-volt) is to look at the front headlights. If the light has a tapered housing with a single wire leading to it, it should be a 6-volt system. If the light has two wires going to the housing and the housing has a flat or rounded back, it should be a 12-volt system.

All Cub and Cub Lo-Boy tractors (except the 184 Lo-Boy) have a generator fitted under their hood. If your tractor is a *not* a 184 and has an alternator fitted, the factory generator has been replaced.

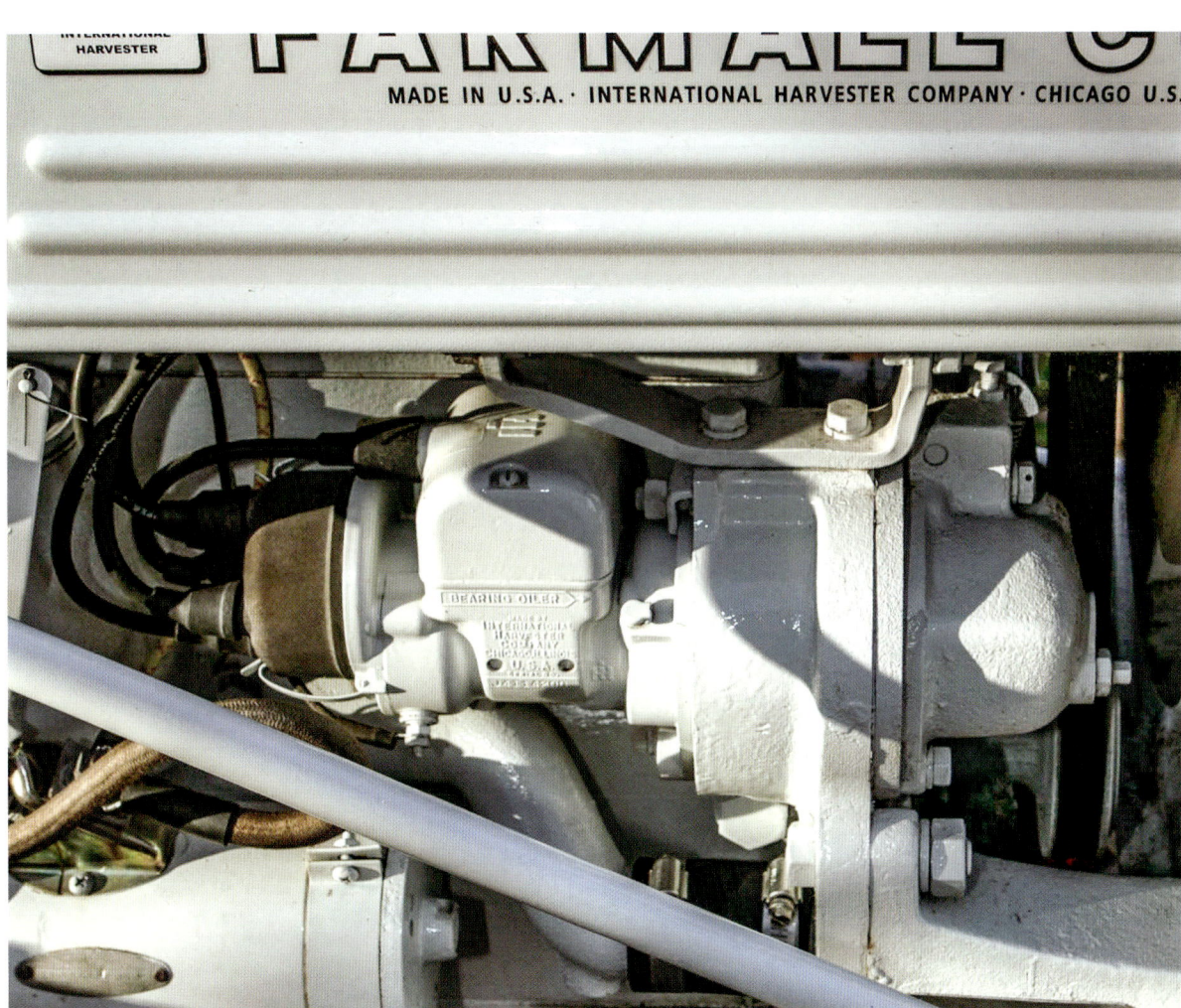

IH MAGNETO IGNITION
IH's very own magneto was used on the Cub. The model J-4 magneto was simple and needed no battery to work. All hand-crank-start Cubs have a magneto as standard factory issue. *Lee Klancher*

Cooling System

The Cub's engine cooling system was a copy of that used by its "big brothers," the Farmall A and B tractors. All three feature a cooling system that uses the thermosyphon principle to cool the engine. Here, coolant (whether it is water or antifreeze) circulates through the engine and radiator by thermosyphon action. As liquid heats, it will expand and move vertically through the system. The liquid cools it as it rises, creating circulation. Steam locomotives are prime examples of this (they use highly pressurized water). As the coolant gets hot, it expands and enters the top tank of the radiator. The cold air blast from the engine-cooling fan cools the liquid, causing it to settle to the bottom of the radiator and then back to the engine block to replace the heated coolant. This explains why Cubs and Cub Lo-Boys do not have engine water pumps. They never did and never will. The radiator was built of a flat-tube design that was protected by the grille and bolted to the steering gear housing base. A gasket sealed the connection between this housing and the radiator. A fan shroud on the radiator increased the engine fan's efficiency greatly, thus reducing the chance of overheating. It is imperative the radiator be checked daily, not only for coolant level, but also for cleanliness. If the fins become bent, they can be carefully straightened with a radiator "brush." *If* the radiator is clogged, have it removed and taken to an IH dealer for service. The use of radiator "stop leaks" is not recommended as they highly reduce the efficiency of the radiator to transfer heat. IH also recommended that the coolant be changed twice yearly. A handy drain plug was located at the bottom of the front steering housing to accomplish this. When refilling the radiator, add coolant until it is visible, below the filler neck. If the coolant level gets too high, the system equalizes the pressure and volume by allowing excess to escape via a relief valve.

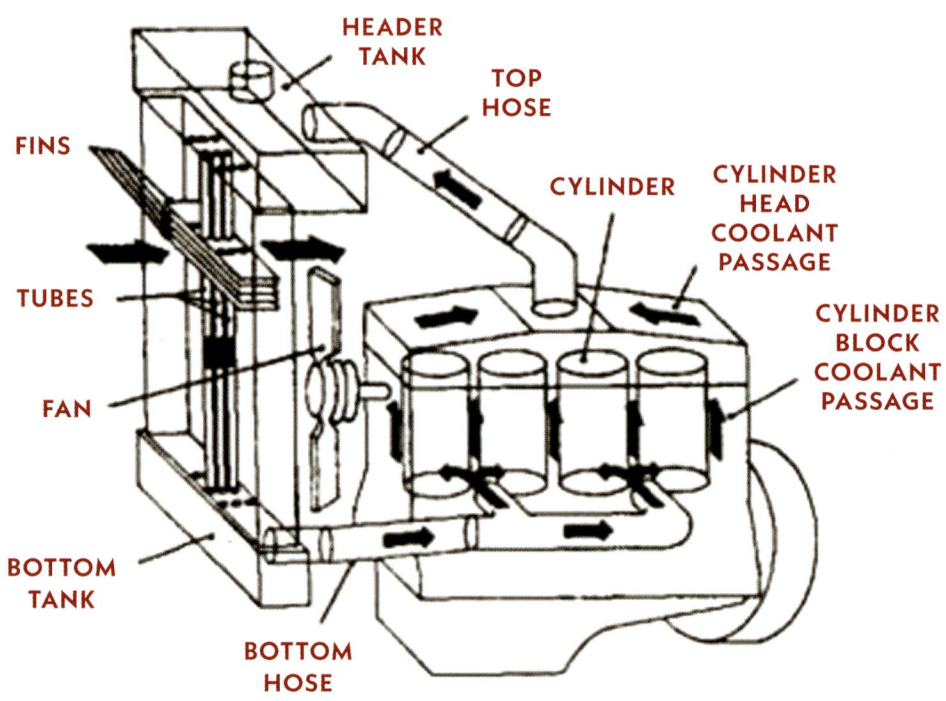

CUB COOLING SYSTEM
The Cub used a thermosyphon cooling system that has no water pump or thermostat. The principle is that hot water rises and flows into the radiator. As the fan blast cools the water, it sinks and flows back to the engine. *Kenneth Updike Collection*

CUB COOLING SYSTEM
The Cub Lo-Boy has the same thermosyphon cooling system as the Farmall Cub. This system is not pressurized. The front bolster acts as the lower radiator tank for the system. *Lee Klancher*

Transmission and Final Drive

The Farmall Cub transmission had three forward and one reverse speed with a bull gear/pinion final drive. The transmission speeds in the Cub were 2 mph for first gear, 3 mph for second gear, and 6 mph in third. The reverse speed was 2 1/4 mph. The transmission was built using sliding spur cut gears, and *was not* synchronized, meaning that downshifting on the go was not possible without loud (and sometimes violent) gear clashing. A foot-operated clutch controls the transmission of engine power not only to the driving wheels but also to the rear power take off/belt pulley (if equipped). This clutch was built as a single-disc, single stage–type clutch. The standard equipment clutch was of the Auburn brand with a Rockford brand being the optional replacement. A pressed graphite bearing was used as the release bearing in the Cub. A true roller bearing (like those on the larger IH tractors) would never be outfitted on the Cub. This stationary release bearing would engage the drive clutch's release levers (sometimes called fingers) to disengage the engine drive clutch. An operator who would "ride the clutch" was one who would be partially depressing the clutch pedal all of the time. This not only rapidly wore out the release bearing, it also did not allow the clutch pressure plate to fully engage the driving disc, allowing the disc to "slip." This slippage of the disc caused tremendous heat to be generated, oftentimes warping the disc or getting the pressure plate hot enough to melt the disc lining. When you smell "burnt clutch" while driving, either someone is riding the clutch, or it is in dire need of replacement/adjustment.

The use of a variable-speed engine governor allowed the operator to get those "in between" speeds for varying crop conditions. The transmission used

FARMALL CUB HIGH-CROP GEAR REDUCTION
The Farmall Cub has massive cast iron gear reduction housings (one for each rear wheel) that not only gear the tractor slower, but provide the "lift" for added ground clearance. *Lee Klancher*

the latest design of spring-loaded rawhide oil seals that kept dirt and dust out. When overhauling a Cub transmission, it is not uncommon to find that today's replacement seals are sometimes one-half of the thickness of the original seals. Vast improvements in sealing material technology have made these parts very compact in thickness. This author highly recommends that when a situation occurs like that listed above, two seals be installed. While this may add to the net cost of the repair, if a single seal is installed and it happens to rest exactly where the old seal did (which probably has worn a slight groove in the shaft), oil will leak past the seal as if nothing were installed at all. IH used over twenty-three ball-and-roller bearings in the Cub to keep friction to a minimum and assure smooth power transmission from the engine to the draw bar. Two distinct shifting levers are used on the Cub and Cub Lo-Boy. The straight lever is for use on all Cubs without a deluxe cushion seat (this has a pipe frame with a small backrest and square seat cushion). An angled lever was used on all of the Cub Lo-Boys and the Cubs with a deluxe seat. The angle lever shifter was also used on all gear drive Cub Cadets. The Cub transmission is filled with 90-weight gear oil lubricant.

The use of large bull pinion final drives gears encased in separate gear housings allowed the cub to have over 20 3/8 inches of crop clearance. This was vital when cultivating high-value vegetable crops. Crop damage could be minimized. The gears in these housings have induction-hardened teeth and ride on tapered roller bearings to take heaviest punishment of tasks. A precision automotive style differential with spiral bevel ring and pinion gears was constantly bathed in oil and was designed to handle large torque shock loads.

CUB LO-BOY WITH BLADE
The Cub Lo-Boy had the same final drive housings as the Cub tractor, but they had been rotated 90 degrees to face forward. By doing so, the tractor's overall height was reduced and the Lo-Boy was born!
Wisconsin Historical Society

Axles, Wheels, and Tires

In the rear, adjustable tread wheel rims mounted on stamped-steel wheels allowed wheel tread settings from 40 inches to 56 inches wide. A set of special 5-inch axle spacers was optionally available for even wider wheel tread applications.

The Cub had as standard equipment a non-adjustable tread width front axle. An adjustable tread axle was optional. This adjustable axle sold for an extra $12 in 1950. To adjust the tread width, the operator moved a pin and bolted clamping blocks in various holes in the front axle tube. These holes allowed the front wheel spacing to be adjusted in 4-inch increments from 40 5/8 inches to 56 5/8 inches in width. IH offered cast iron weights in 26-lb size to fit the front wheels and in 145-lb size for the rear wheels. Both styles of weights were bolted to the wheel disc via two bolts, and additional weights could be "stack bolted" to optimize tractor ballasting requirements. The front wheel weights were also used as the rear

FARMALL CUB WHEELS
Both the front and rear wheels of this Cub have cast iron weights bolted to them. The extra weights add stability to the tractor and they aid in getting better traction, too. *Lee Klancher*

wheel weights on the Cub Cadet garden tractors. Today, these cast weights can be scarce to find and are highly sought after by collectors.

IH offered a choice of tire sizes for the Cub and Cub Lo-Boy. They did not offer the choice of having a specific tire brand. IH did use major name tires like Goodyear, Goodrich, and Firestone during the tractor's production run. If the buyer wanted a specific tire brand, they could be exchanged locally by the dealer or buyer of the tractor. IH offered four rear tire size choices on the Cub and Cub Lo-Boy. The options were: an 8.3-24-inch, 4-ply ag lug tire (this was standard equipment); an 8.3-24-inch, 4-ply industrial; a 9.5-24-inch, 4-ply ag lug tire; and a 9.5-24-inch industrial tire. The industrial tire has a diamond tread lug pattern.

The front wheels were a 4.00-12-inch ribbed tire. IH never used car tires or sizes other than those listed on the Cub and Cub Lo-Boy. The numbered Lo-Boys (154, 185, 184) had a completely different set of tire-size choices.

CUB LO-BOY WHEELS
This Cub Lo-Boy does not have any added wheel weights, only the stock IH Red–painted wheel rim discs. The rear wheel rim is painted silver. *Lee Klancher*

Touch Control Hydraulic Lift

An optional feature that was highly sold by IH dealers was the Touch Control hydraulic lifting system. This, too, was a scaled-down version of that found on the Super A or Farmall C.

The first Cubs made did not have the Touch Control offered as factory install as the system was not perfected yet. IH did however machine the castings to allow the addition of the Touch Control system at a later date by an IH dealer if the customer chose to add this. The major difference was that the Cub has a single control lever, single rockshaft, and single hydraulic cylinder. The Touch Control system offers two-way control of mounted implements. This was especially handy to put down pressure on front-, center-, or rear-mounted grading blades. An adjustable indicator on the Touch Control lever quadrant is handy to use when a uniform implement working depth is required. The Touch Control unit mounts directly behind the engine under the gas tank. It looks like a large block with rotating lever arms on each side. A rear-mounted rockshaft can be added to tractors whose implements require this. With Touch Control hydraulics, the operator can effortlessly lift heavy implements with the touch of a finger. An engine-driven hydraulic pump located on the left-hand side of the engine powered the Touch Control unit. This pump ran whenever the engine did as it was driven via the timing gears on the engine. A mesh screen inside the Touch Control block acted as the filter for the system. The Touch Control unit was fully self-contained with its own lubricant called Touch Control fluid. This oil has been greatly refined over the years and now, as previously noted, is known as Hy-Tran Ultraction. While this Touch Control hydraulic lift was used exclusively for implements when first installed on the Cub, it would later be the lifting power source for the rear hitching system called the Fast Hitch.

TOUCH CONTROL BLOCK
The IH-built Touch Control block fitted on the Cub. IH used two different blocks, and you need to know the block part number (here, 354383R1) to get the correct repair kit. *Lee Klancher*

LO-BOY HYDRAULIC LIFT
The Cub Lo-Boy shared the same hydraulic lifting system found on the Farmall Cub. Moving heavy snow with a front-mounted blade is easy on the operator when the blade is lifted by moving a single hydraulic control lever. *Wisconsin Historical Society*

The Hydraulic lifting unit on the Cub tractor was called the Touch Control system. This was controlled by levers near the steering tower. By moving a lever, a hydraulic cylinder connected to a rockshaft moved a set of arms that were externally mounted from the Touch Control unit. The Touch Control unit was located under the gas tank. The gear-type hydraulic pump that powered this unit was driven by gears inside the engine's front cover. Two steel lines from the Touch Control unit were connected between the pump and the control block. IH used o-rings and gaskets to seal the line. The pump itself has a single o-ring that keeps hydraulic oil from escaping the pump and entering the crankcase. This o-ring on the pump is known to fail with age or a pressure spike in the system. This allows the hydraulic oil in the Touch Control block to quickly flow into the engine crankcase, overfilling it.

If your engine oil level is overly full or "grew" overnight, chances are the o-ring on the pump has failed. By unbolting the steel lines and then removing the pump and replacing the o-ring, the problem is solved. The Touch Control unit has an internal brass mesh filter screen that can be removed for cleaning. A gasket/seal overhaul kit is offered by IH to repair the Touch Control block. You do need to know the casting number on the block to get the correct gasket/seal kit. The tractor's serial number has nothing to do with which block is on the tractor.

The Cub's hydraulic Touch Control system was not ready for production when the tractor was introduced. The system can be readily added (or deleted) to any year Cub that is not fitted with it.

CHAPTER 4

How to Buy a Cub

Buying a Farmall Cub is an adventure! As with any old tractor, it will likely show evidence of many different mechanics who have worked on the tractor over the years—each with their own style and level of attention to detail. The lower the price, the more imperfections you should expect to see. Every used tractor will be a "fixer-upper," and it's wise for buyers to begin the process with a realistic assessment of what they can afford: Are you looking for a tractor that needs major work, or are you willing to pay top dollar for a tractor with a complete, recent overhaul?

Unlike when purchasing a car, you can't rely on VIN numbers and maintenance records to determine the tractor's mechanical soundness. You'll have to conduct your research the old-fashioned way, by thoroughly inspecting the tractor yourself. Here's a quick guide to the questions I ask myself every time I inspect a tractor for purchase. Following these steps can help you to find a good buy of your own.

Visual Inspection

1. How are the tires and rims?
Look for matching tires, correctly sized, and not too worn. Replacing a tire is straightforward, but used tractor tires can be hard to find and new ones are very expensive. If the tires will need to be replaced, be sure you calculate this into the price you're willing to offer.

2. Is the sheet metal in good condition?
It's common for owners to cut the hood to make room for an alternator during a 12-volt conversion. While a tractor will operate just fine with less-than-perfect sheet metal, if you are aiming for a full restoration you'll want to consider this.

3. Are there any leaking fluids?
Inspect both the tractor itself and the ground where it has been sitting. Leaking oil at the rear main seal is a red flag—replacing this seal is a very involved repair that requires splitting the tractor in half. Pull the dipstick and look for water in the oil (it's polite to bring your own rag). Water in the oil could indicate a blown head gasket or a cracked head/block.

4. Is the wiring in good shape?
Look over the tractor's wiring, checking for wires that are frayed or chewed on. Extra wires hanging loose are a red flag, too. If the tractor has had a 12-volt conversion, look for a neat job with wires in good condition.

5. How is the general upkeep of the tractor?
A good clue to the tractor's condition is where it has been stored—especially in colder climates. Cosmetic details—like a good paint job and proper decals—often point to a tractor that has been well cared for on the mechanical side as well.

6. Are any castings broken or cracked?
Look for freeze damage and stress cracks, particularly on the engine block and head, on the front bolster, on the final drive housing, and on the implement mounting holes. Cracks to the engine block and head can be effectively repaired (although steer clear of poorly-done repairs, which can fail and cause extensive damage to the entire engine). A tractor with a cracked final drive housing or implement mounting holes is an easy fix, as replacement parts are plentiful and the swap is straightforward. Damage to the front bolster is very difficult to correct. There is too much torque involved for a simple weld to hold, and replacing the bolster is a very extensive project. Unless the tractor has sentimental value, I would recommend walking away from a tractor in need of a front bolster repair.

Mechanical Inspection

Ask the seller if it is okay for you to start the tractor and drive it around. If the seller agrees to a test drive, be sure you ask about the condition of the brakes—that's not something you want to be surprised by!

As you start the tractor, pay attention to these things:

1. Does the charging system work?
The tractor should start smoothly with a quick pull of the starter ring. Look for the amp gauge to quickly rise and hold steady at a positive charge. If the tractor requires a hand-crank to start, or if the amp gauge is negative or neutral, you likely have a problem with the charging system. Charging system problems are rather easy to correct on a Cub—if the tractor's price is low enough, don't be scared away!

2. Does the engine start smoothly? Does it smoke or miss?
A bit of dark exhaust is entirely normal when an engine is first started, but if it is still blowing smoke after a few seconds you've got a problem on your hands. If it misses, that's also a problem. Listen to the tractor run for a bit before you attempt to shift it into gear.

Driving Inspection

1. Depress the clutch and gently shift the tractor into first gear.
Release the clutch slowly, and make sure the brakes work before you really take off! If the brakes work, drive around a bit and attempt to cycle through all the

gears. Look for a clutch that operates smoothly, with no grinding gears. Compared to other tractors, a Cub clutch is a particularly difficult repair. Unless you are getting a great deal and are up for a big project, steer clear of a Cub with a clutch problem.

2. Test the brakes.
If there is some play, don't worry too much. This is an easy repair. If the brakes do not work at all, consider the expense of a complete brake repair.

3. If the tractor is equipped with Touch Control Hydraulics, give them a good test.
The hydraulic system is the other component of the tractor that I think is the most complicated to fix (perhaps even equal to the clutch in difficulty). If the hydraulic system needs work, beware.

For a Non-Running Tractor
1. Ask "When did it last run?" and "When it was parked, was it all working or was it a project tractor?"
A tractor that was in excellent shape and has been sitting for less than a year might only need a charged-up battery. On the other hand, a tractor that was purchased with problems and has sat for years since then is a much riskier buy.

2. Sometimes the seller isn't the person who worked on the tractor last.
This is most commonly the case with an estate sale. If the seller doesn't know much about the previous condition of the tractor, a good clue is the tractor's storage location. One that has been stored indoors is likely in better condition than one from the fencerow.

3. Try to roll the engine over with the hand crank – is it stuck?
A tractor that is seized will require a lot more work than one with a free engine.

4. Consider the overall condition of the tractor.
The condition of the paint and tires is a good clue of how far gone it might be. Note that a set of new tires can cost you as much as the tractor—if the tires are ruined or unusable, check the price of a new set before you buy.

5. Carefully inspect the block.
Look for cracks and welds or water in the oil.

Closing the Deal
1. Ask if the tractor comes with a manual or any equipment.
It's not uncommon for a seller to gesture toward a fencerow of junk and say, "Oh yeah, all that stuff can go with it, too." I've snagged plenty of useful and rare attachments this way, free for the asking!

2. Ask "Are you aware of any issues that will need immediate attention?"
This is the most useful question I ask—it's a good way to catch anything you might have missed in your inspection.

3. Agree on a fair price.
Take into consideration the expense of the repairs that you intend to do—but don't nitpick a lower price for every minor imperfection. If you can't agree with the seller on a price, ask if you can leave your phone number just in case a better offer doesn't come along.

CHAPTER 5

Cub Implement Guide

This is a partial list of implements that IH built and designed for the Cub and Cub Lo-Boy series tractors. Only those implements that IH offered through its dealer distribution channel are listed.

IH offered Cub-specific implements that were sized for the Cub tractor to handle, and not just a cut-off chunk of a larger implement. IH built plows and cultivators for the Cub at its Canton, Illinois, factory. The IH factory in Hamilton, Ontario, Canada produced disc harrows, and the Memphis, Tennessee, factory owned by IH produced corn and cotton planters for the Cub.

The Cub was the smallest model in IH's line of Culti-Vision tractors. Culti-Vision was a system that placed the driver over one row of crop, with a good view for cultivating. This was done by placing the operator platform offset from the engine and driveline. Culti-Vision offered an unmatched line of sight when working with grown/growing plants to prevent tractor-versus-plant damage. Most of the Cub and Cub Lo-Boy implements were designed to be mid-mounted on the tractor to give the driver the full benefit of the Culti-Vision design.

The implements to fit the Cub and Cub Lo-Boy tractors were identified with a "Cub," "F," or "L" prefix. The Cub prefix meant it fit the Cub tractor. The "F" prefix meant the implement was Fast Hitch and the "L" prefix meant that the implement was designed specifically for the Lo-Boy tractor model. Using this information an LF38 disc harrow is made for a Lo-Boy tractor fitted with a Fast Hitch. The Cub-F153 disc plow is meant to fit a Cub with a Fast Hitch. The specific list of IH-made implements is on the next page.

Cub Implement	Description/Notes
A-Cub-33	Bean Harvester, one row
Cub-144 cultivator	One-row cultivator
Cub-252 cultivator	Two-row cultivator
Cub-447 cultivator	Four-row cultivator
Cub-3 spring-tooth field cultivator	
Cub leveling and grader blade	Can be front or mid-mounted
Cub-16 middlebuster	
Cub-22 sickle mower	4 1/2-foot-wide cut
Cub-23-A	Tandem disc harrow
Cub-38 disc harrow	
F-38 disc harrow	Fast Hitch disc harrow
LF38 disc harrow	Lo-Boy Fast Hitch disc harrow
Cub 135 planter	One-row vegetable planter
Cub 170 planter	One-row planter
Cub 171 planter	One-row backland cotton planter
Cub 172 planter	One-row runner cotton planter
173-A backland planter	
174-A runner planter	
Cub-201	Two-row drill-type planter
Cub 435 planter	Four-row vegetable planter
Cub 474 planter	Four-row vegetable planter
Cub-193 moldboard plow	
L-F194 moldboard plow	Lo-Boy Fast Hitch moldboard plow
L-F11 moldboard plow	Lo-Boy Fast Hitch moldboard plow
Cub-189	One-furrow, 12-inch, two-way moldboard plow
Cub-193	One-furrow, 12-inch, one-way moldboard plow
Cub-151	One-furrow disc plow
Cub-152	One-furrow disc plow with Timken roller or Chilled-bearing
F-153 disc plow	Fast Hitch disc plow
Cub-12-D	2-disc harrow plow
Cub-6 tool bar carrier	Rear mounted
Cub 2-wheeled farm trailer	
54A blade	
L54 blade	54-inch blade for Lo-Boy
Cub LF-3 spring-tooth harrow	Lo-Boy Fast Hitch spring-tooth harrow
Cub LF-4 peg-tooth harrow	Lo-Boy Fast Hitch peg tooth harrow
Cub LF-11 rotary hoe	Lo-Boy Fast Hitch rotary hoe
L-22 sickle mower	4 1/2-foot-wide cut
Lo-Boy sickle mower	
1000 hydraulic front-end loader	

The caption information for these historical black and white images was sourced from original IH literature.

ONE-WAY MOLDBOARD PLOW

Take a look at the one-way moldboard plow. The plow does a thorough job of turning soil under most conditions. It is as simple and rugged as a walking plow, and was offered at a low cost. It cuts a 10-inch furrow—down to 8-inches deep—at a rate of 3 1/2 acres per day. The strong, heat-treated, high-carbon plow beam is quick-connect attached to the reversed tractor drawbar. Two cushion springs absorb shocks and allow the bottom to dodge hidden rocks without damage to the share. A wide variety of bottoms were offered to meet the soil requirements of the buyer's locality. Depth is controlled by a hand lever that lowers and raises the plow hitch. The plow can be easily lowered or raised, either by hydraulic Touch Control or a hand lever. *Kenneth Updike Collection*

TWO-WAY MOLDBOARD PLOW

This is essentially the same plow as the one-way except it has an additional left-hand bottom. By having both a right- and left-hand bottom, all furrow slices can be turned in the same direction, meaning no dead or back furrows. The plow is ideal for contour farming—each furrow slice can be turned uphill to form a small water- and soil-holding terrace. The plow excels at building and maintaining terraces. For irrigated fields, the plow is almost a necessity to keeping the fields level. A latching mechanism controls the dropping of either bottom. *Kenneth Updike Collection*

CUB DISC PLOW

This is the right tool for plowing if you have hard and dry ground, sticky and abrasive soil, or stony and root infested fields. This one-furrow disc plow is also first-rate for building and maintaining terraces. The 26-inch disc is heat-treated to take hard wear, and can cut a 10-inch furrow. The furrow wheel links to the tractor's steering, yielding a full-width furrow when plowing on curves and sloping land. An easy-to-reach lever levels the plow. A special hitching device—the equalizing drawbar—lets the plow move from side to side without shifting the draft from the center of the tractor. This makes steering easier and lends to a better plowing job. *Kenneth Updike Collection*

TANDEM DISC HARROW

No patch is too small and no corner is too tight for this 4-foot tandem McCormick disc harrow. The implement can be maneuvered close to fences, vines, and trees to make productive use of every possible square foot of ground. The heavy-gauge steel discs—your choice of 14- or 16-inch size—are vertically ground and heat-treated to provide years of dependable service. A simple cross-draft construction forces the rear gang to follow the front, reducing skidding on corners or contours. The operator can back into tight corners without jackknifing. From the tractor seat, the sections can be set to five different angles, providing just the implement needed to create a fine seedbed. *Kenneth Updike Collection*

SINGLE-CUT 5-FOOT DISC HARROW

This McCormick disc is highly maneuverable and will go almost anywhere a Farmall Cub will go: small fields, irregular plots, gardens, orchards, or truck patches. Though it makes a single cut instead of a double cut like the tandem, it cuts 5 feet instead of 4, and covers about 1 3/4 acres per hour. Both 16- and 18-inch discs were offered. A screw-type angling device sets the angle of the discs. The implement is sturdily constructed, penetrates the soil deeply, and does a top-notch job even in fairly hard ground or trashy fields. *Kenneth Updike Collection*

PEG-TOOTH HARROW

For the final pulverizing and smoothing of the seedbed before planting—as well as early cultivation of select crops—this McCormick peg-tooth harrow is highly useful. The Farmall Cub easily handles two of these 4-foot-9-inch sections, and covers about 3 1/2 acres per hour. The lever sets the angle of the teeth with ease, and folds back when not in use so as to take less space in storage. The ends are closed to keep from hanging on trees or fences as well as to prevent damage. Sturdy cross-bars distribute the strain yet provide flexibility for uneven ground. *Kenneth Updike Collection*

WEEDER-MULCHER
Here's a handy implement for controlling weeds in crops like corn, cotton, beans, peanuts, and other sturdy row-crops. The springy, pencil-point teeth get right down into the soil and dig out pesty little weeds that surround the plant without damaging the deeper-rooted crop. At the same time, they break the sun-baked crust and mulch the soil, restoring the ground to a healthy growing condition. It's good to use after every rain—or every seven to ten days—while the plants are still small. The 10-foot McCormick weeder-mulcher can cover three rows at a time of wider-spaced crops like corn or cotton, and cover about 3 1/2 acres per hour. *Kenneth Updike Collection*

SOIL PULVERIZER
The McCormick pulverizer firms the soil and flattens out air spaces to reduce blowing, conserve moisture, prevent winter-killing. It's just the thing for working in wheat, oats, alfalfa, or other such crops. The implement helps increase germination, and if used shortly after drilling or seeding, it well repays its cost in seed saved. It can also be used for seedbed preparation in corn or other row-crops. A gross seeding attachment was offered. The Farmall Cub readily handles a 7-foot pulverizer at 3 miles per hour and covers about 2 1/2 acres per hour. *Kenneth Updike Collection*

HAY LOADER

The McCormick cylinder-rake loader provides an easy way to load hay, which saves the farmer a lot of sweat and pitchfork swinging. It makes fast, clean work of getting hay onto a rack. The implement picks up the hay without bunching or tearing, and handles it gently to keep from shattering the leaves—an important feature with clover or alfalfa. Hay can be loaded from either windrows or swaths with the 6-foot raking width. It does a good job on hillsides, rough ground, and almost anywhere you want to use it. A different loader was offered for handling green crops such as peas and beans for the cannery. *Kenneth Updike Collection*

SIDE-DELIVERY RAKE

The McCormick side-delivery rake teams up with the Farmall Cub to do a good, clean, and fast raking job. The machine makes loose, fluffy windrows with the majority of the leaves inside and the stems outside. The implement ensures that hay cures quickly and evenly, with less damage by the sun and rain. Hay stays bright and green, with less shattering of the nutritious leaves. Since the entire reel unit floats up and down over rough spots, it does a clean job of picking up the hay. The V-type steel tractor hitch makes it trail closely and accurately, allowing the operator to get into tight corners and navigate close to trees and fences to save every bit of hay. The Farmall Cub also teams up well with a dump rake. *Kenneth Updike Collection*

4 1/2-FOOT SICKLE BAR MOWER

Take a look at the fellow on the left. He's cutting weeds and grass along his driveway—with ease and speed—using the side-mounted 4 1/2-foot mower. This handy machine mows grass and weeds on both the yard and farm. It's just the implement needed for mowing hay in small fields. A good view of the cutter bar and field ahead makes it easy to reach into corners and mow alongside fences, trees, and shrubs. Touch Control raises and lowers the cutter bar (a hand lever control was available) to easily avoid rocks or stumps. It's driven from the power take-off through a "V" belt, which gives quiet, smooth operation—no gears to wear or break. A choice of cutter bars was offered for any crop.
Kenneth Updike Collection

TWO-ROW BEET AND BEAN CULTIVATOR

This implement is ideal for beets and beans, of course. But that's not all. The cultivator suits many other crops planted in narrow rows. It teams up with the Cub vegetable planters and can be adjusted quickly to handle four 12-inch, three 16- or 18-inch, or two 22- or 28-inch rows when the Cub is equipped with an adjustable front axle. Cubs equipped with a regular front axle can handle three 12-inch or two 16- to 20-inch rows using the cultivator. A choice of ground tools was offered to accommodate varied soil and crop conditions, including diamond points, duckfoots, knife weeders, disc weeders, and deer tongues. The attachment process is again one of ease and speed. The cultivator can be slipped onto the Universal Mounting Frame and be in the field in less time than it takes to hook up a team of horses. Cub cultivators handle more acres in less time, making them a valuable tool for the part-time farmer.
Kenneth Updike Collection

FOUR-ROW VEGETABLE CULTIVATOR

This cultivator is an indispensable helper for growing vegetables. It can be used on any system of row spacing set by the Cub front-mounted vegetable planter. With the easy-to-see row guide, it is easy keep right on the row, work all the soil, and get the weeds close to the plants. Each cultivating unit is free to follow the soil terrain as a gauge wheel controls the depth. Jockey bars link the units together to hold them accurately on the rows. The result is a perfect job with fewer plants covered up. When moving from one type of vegetable to another, changing to new row spacing is simple and easy: just loosen the cap, slip the unit into the desired position, and tighten. Notice that the operator has left the Cub 474 planter hoppers and drive mechanism mounted on the tractor while cultivating, saving time in changeover. *Kenneth Updike Collection*

ONE-ROW CORN AND COTTON CULTIVATOR

This popular, all-around cultivator fits the small, diversified farm. Mounted on a Farmall Cub with an adjustable front axle, the implement can work 36- to 56-inch rows, cultivate clean and fast, and cover up to 12 acres per day. A wide choice of ground tools was offered. Regular equipment includes parallel floating shields (except with spring teeth). There are seven tool combinations, including high-speed sweeps for clean, shallow cultivations, double-point shovels for pulverizing and deep-working, and spring teeth for rough, stony ground or for the removal of pest weeds like bindweed and quack. Special equipment includes a fertilizer side-dressing unit, potato hiller, covered shields, and jockey arch for the rear section. The same fertilizer unit can be used for the single-row planter and cultivator. The cultivator units consist of just the ground tools and tool bars. They are easy to attach and detach, and the adjustment of the tools is never disturbed. *Kenneth Updike Collection*

Model 1000 "One-Arm" Loader

No book on the Cub and Cub Lo-Boy tractors would be complete without mention of (probably) the most sought-after attachment for these tractors, the Model 1000 hydraulic front-end loader. This is commonly called the one-arm loader among IH enthusiasts. This all-hydraulic loader was painted IH White and was made from 1962 to 1975. It actually was oversized for the Cub, with its loader frame attached to both the tractor's rear axle and the front steering casting. A separate control valve operated the loader using the tractor's hydraulic pump as its power source. A Cub with a front-end loader attachment can be a very useful machine, but the front axle, steering gearbox, and especially the spindles and narrow wheel rims used on the Cub greatly hindered the performance of this loader. The 1000 loader could easily overload the Cub's front axle and possibly cause the axle to break. The 1000 loader could be installed on either the Cub or Cub Lo-Boy tractor. The nickname "one-arm loader" was given to the 1000 as it has only one loader-lifting boom arm. Most front-end loaders have two parallel lifting arms. Another unique feature is the use of a single hydraulic cylinder to tilt the bucket. The only major design change IH made to the 1000 loader was in the bucket tilt cylinder design. The earlier style (1962–1965) used a metal pipe attached to the cylinder barrel. The post-1965-built models used a cylinder with 90-degree elbows on the ends of the hydraulic hoses instead of the pipe.

Note that the model 1050A loader differed from the 1000 loader in that the 1050A was designed to fit the Cub Lo-Boy 154. The 1050A was built from 1970 to 1973 and used two loader-lifting arms and dual hydraulic cylinders to tilt the front bucket. The Model 1000 loader also fit the Farmall 140. When considering purchase of a 1000 loader, inspect closely for frame modifications. The rear of the frame was often torched off when mounted on a Lo-Boy, rendering it useless for any tractor except a Lo-Boy. When mounted on a 140 with an industrial axle, the front mount was modified, making it unusable on a Lo-Boy.

1957 CUB LO-BOY WITH 1000 LOADER
The 1957 model year Cub Lo-Boy here is fitted with the IH Model 1000 one-arm front loader. The low profile of the Lo-Boy tractor has a lower center of gravity that made this a very stable loader tractor unit. *Lee Klancher*

1963 CUB WITH 1000 LOADER
The 1000 loader uses the tractor's hydraulics to power its boom- and bucket-operating hydraulic cylinders. *Lee Klancher*

Model 154, 185, 184 Lo-Boy Implements

The numbered series Lo-Boy tractor have their own unique implements due to their frame and lifting mechanism. This list includes the later models, the 154, 184, and 185. The list of implements IH offered for these models is below:

Model 54 blade	54-inch-wide front grader blade
Model 3142 mower	42-inch-wide-cut, two-blade rough-cut mower
Model 3160 mower	60-inch-wide-cut, three-blade finish mower; fits 154
Model 3260 under-mounted	60-inch-cutting-width rotary mower
Model 110 disc harrow	52-inch-wide, three-point hitch tandem disc harrow
Model 1050A loader	Hydraulic front-end-mounted loader
48-inch mower	48-inch-cut, three-blade finish mower
60-inch mower	60-inch-cut, three-blade finish mower
Model 222 mower	60-inch-cut, side-mounted sickle bar mower
Model 3160A mower	60-inch-cut, three-blade finish mower; fits 184
Model 3620 mower	60-inch-cut, three-blade finish mower
Model 310 moldboard plow	three-point hitch, 12-inch-cut moldboard plow
Model 15 rotary tiller	three-point hitch 42-inch-wide rotary tiller
Model 50 snow thrower	Front-mounted snow thrower
Model 10 rear blade	6-foot-wide, three-point-mounted rear blade
Model 110 tandem disc harrow	
Model 1200	Tow-type, rear-tilting flatbed trailer
Model 1050A loader	Loader for Cub 154

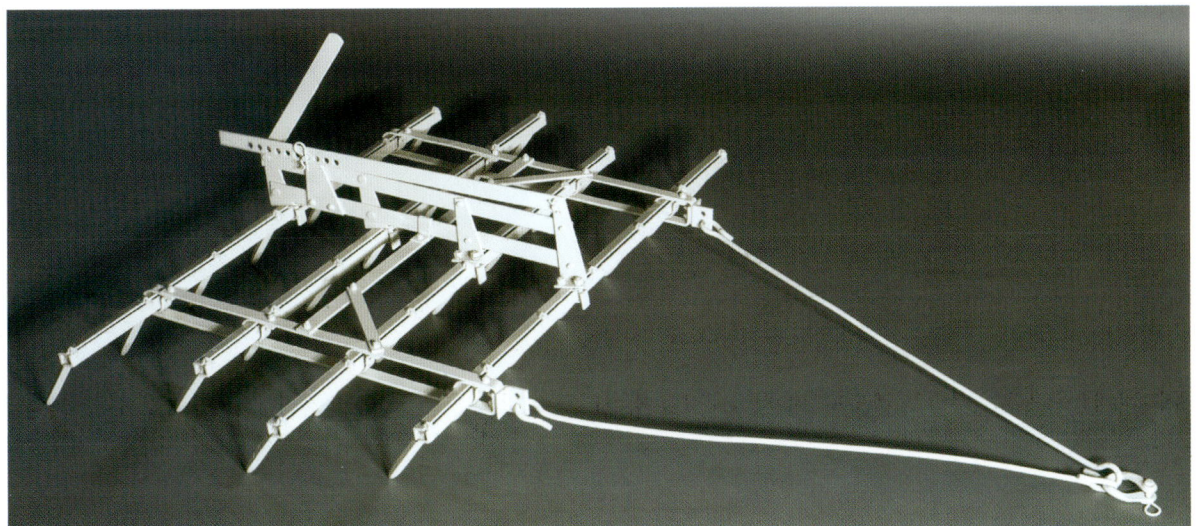

SPIKE-TOOTH FOUR-BAR HARROW
This pull-type harrow has heavy metal spikes (they are replaceable) affixed to four angle iron bars. When pulled, the spikes break up the soil clods. The adjustable handle determines the working depth of the harrow. *Wisconsin Historical Society*

Some 1950 Implement Pricing

Using the year 1950 in our example (year of the white demo Cub), here are some list prices of Cub tractor implements that IH offered:

Cub #144 cultivator	$75
Cub three-spring-tooth field cultivator	$61
Cub #23-A tandem disc harrow	$135
Cub #22 4 1/2-foot sickle mower	$97
Cub #435 four-row vegetable planter with markers	$179
Cub grading and leveling blade	$35
Cub #6 rear tool bar with single-disc harrow gangs	$108
Cub #151 one-furrow disc plow	$128
Cub #193 12-inch-wide-cut moldboard plow	$59
Farmall Cub tractor with hydraulics, belt pulley and PTO, deluxe seat, muffler, and battery ignition	$867

Fast Hitch Implements

In 1955, when IH introduced its new Fast Hitch system of attaching rear-mounted implements, the Cub was not left out. The Farmalls of the time (300 & 400) used a two-prong/socket implement mounting setup. This would not fit on the Cub or Cub Lo-Boy, so IH engineers devised a revised Fast Hitch that would fit these two tractors *only*! The Fast Hitch IH used the same size prong and socket as the large Farmalls, except that instead of having two of each, the Cub and Cub Lo-Boy were fitted with a *single* prong/socket Fast Hitch.

This proved ideal for IH in that *now*, the Fast Hitch implements for the Cub and Cub Lo-Boy were exactly sized to fit the tractor's Fast Hitch. Some crafty blacksmiths could work their magic to make an implement (usually one that was too big!) to fit the Fast Hitch, but that was very much the exception.

IH continued to offer the Cub and Cub Lo-Boy (until its retirement) with a Fast Hitch option. A Category 0 or Category 1, three-point hitch was never offered by IH for *any* of the Cub series tractors.

For IH, the Fast Hitch change to the implements was simple for production and as time passed on, more used Fast Hitch implements became available in the marketplace.

To identify an implement that was Fast Hitch compatible (and made by IH), a letter F was usually added to the implement model number.

IH FAST HITCH DISC HARROW
This harrow would attach to tractors equipped with a rear Fast Hitch. The open frames on top of the harrow were used to place ballast blocks on to aid in soil penetration. *Wisconsin Historical Society*

PLATFORM CARRIER
Large wooden planks form the body of this custom-made platform carrier. The platform carrier was compared to having an extra loader on the tractor. *Lee Klancher*

PLATFORM CARRIER HITCH
Close-up view of the custom-made platform carrier's Fast Hitch. *Lee Klancher*

IH 188 FAST HITCH PLANTER UNIT
This IH #188 plate-type planter unit has a Fast Hitch mounting to allow it to fit on the Cub or Cub Lo-Boy. One row of seeds is placed with this planter. *Lee Klancher*

CLOSE-UP OF #188 PLANTER HITCH
This view shows the IH Fast Hitch tong and socket, along with the related hitch linkage. The Fast Hitch was an exclusive IH item that (sadly) was never licensed to others. *Lee Klancher*

#188 PLANTER
This IH #188 planter unit fitted to the Cub Lo-Boy via the rear Fast Hitch. The Fast Hitch lets you change implements quickly. *Lee Klancher*

SEARS-DAVID BRADLEY SCOOP
The David Bradley Company made a rear scoop to fit the Cub Lo-Boy with a Fast Hitch. This photo shows the scoop and the colorful decal on the rear of it. *Lee Klancher*

1957 CUB LO-BOY WITH FLAIL MOWER
A Fast Hitch-mounted flail mower on this Cub Lo-Boy can make fast work out of grass cutting or the taming of light brush. The flail blades on the mower pivot back when an obstruction may be hit. *Lee Klancher*

1956 FARMALL CUB WITH DIGGER
The solid steel square beam of this digger is needed to handle the draft load from the five coil spring digger teeth. The IH red frame and IH blue teeth are painted in the correct color scheme for the time era. The Fast Hitch makes changing from this implement to another fast and easy. *Lee Klancher*

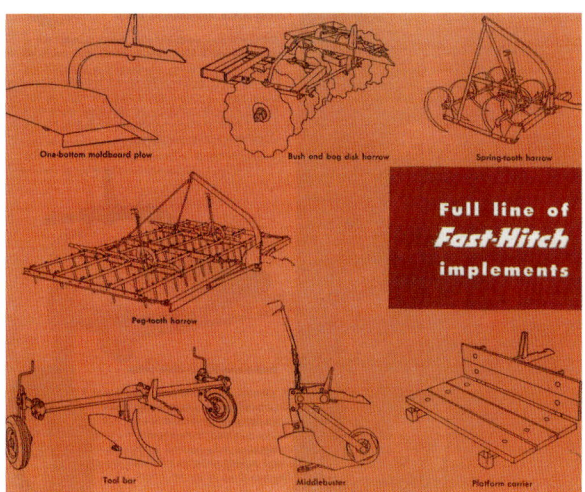

FAST HITCH IMPLEMENTS
The moldboard plow, disc harrow and platform carrier were popular Fast Hitch implements.
Kenneth Updike Collection

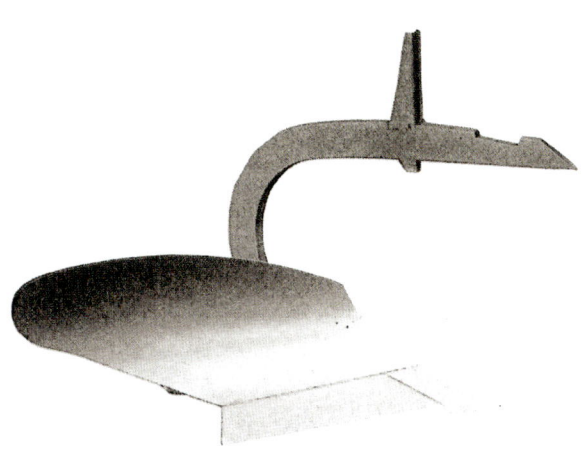

FAST HITCH MOLDBOARD PLOW
This image displays a single-bottom Fast Hitch moldboard plow to fit a Cub tractor equipped with an IH Fast Hitch. *Kenneth Updike Collection*

FAST-HITCH MIDDLEBUSTER
Uniform work can be accomplished on any busting or re-busting job with this husky middlebuster. It is equipped with a 10-inch bottom. The high bail and vertical beam provide plenty of clearance to stay clear of high stubble and stalks. A rolling coulter keeps the middlebuster from fouling in trashy conditions. A gauge wheel replaces the coulter to regulate depth in average conditions.
Kenneth Updike Collection

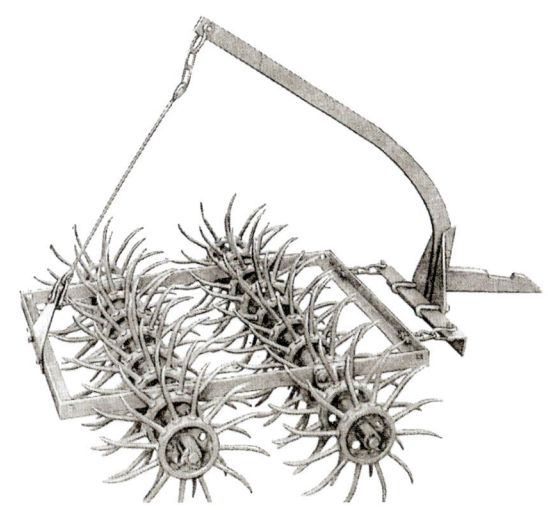

FAST HITCH ROTARY HOE
Use this rotary hoe for fast, economical first cultivations. Knock out weeds before they get a head start. Pulverize hard and caked soil so that sprouting crops can break through easily. This Fast Hitch rotary hoe is 50-inches wide, allowing coverage of 20 acres or more per day with the Cub. The front and rear spoke wheels are staggered to work the ground every 2 3/4 inches. The spokes have sharp, drawn points for easy penetration and are made of extra-strong forged steel to withstand the shock of working in stony fields.
Kenneth Updike Collection

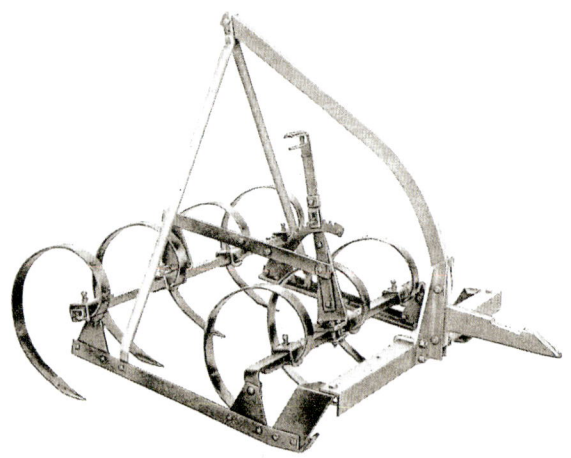

FAST HITCH SPRING-TOOTH HARROW
This harrow works a strip 50-inches wide. It can stir the soil to a depth of 6 inches. It is similar to the popular McCormick No. 2 spring-tooth harrow, but is adapted to Fast Hitch for easy lifting and transporting. The rugged frame is designed to withstand the strains of high-speed tractor farming. An adjusting lever is used to set teeth at proper angle. Wide selections of teeth and points are available. Five and seven-tooth models were offered. *Kenneth Updike Collection*

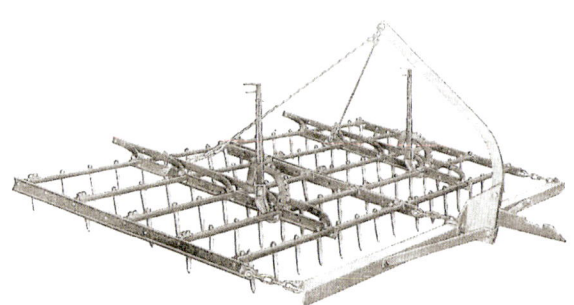

FAST HITCH PEG TOOTH HARROW
The harrow is composed of two 5-foot sections similar to other popular McCormick peg-tooth harrows, but is adapted to Fast Hitch for easy lifting and transporting. Teeth work the soil at 1 1/2-inch intervals. The ground-working width is 122 inches. This harrow has plenty of strength for high-speed tillage work. *Kenneth Updike Collection*

CUB W/ FAST HITCH DISC HARROW
This image displays a rear view of a Farmall Cub tractor with Fast Hitch disc harrow in tow. The rear belt pulley has been removed and replaced by a tubular metal guard. *Kenneth Updike Collection*

FAST HITCH BUSH AND BOG DISC HARROW
The Fast Hitch single-gang disc harrow can be obtained in 4- or 5-foot-1-inch sizes. It can be equipped with 16-, 18-, or 20-inch plain or notched discs. Disc spacing is 6 5/8 inches. The gang angle can be set at any of three positions. Outer ends of gangs are individually adjusted for uniform penetration by means of threaded rods. Individually adjustable full-blade scrapers keep discs clean. Weight boxes are provided, should additional weight be needed for penetration. *Kenneth Updike Collection*

Three-Point Hitch Cub Implements

The arrival of the numbered Cub Lo-Boys (154, 185, 184) had these three tractors fitted with an optional rear three-point hitch and *not* a Fast Hitch. This was a Category 1 three-point hitch in size. The three-point hitch implements are *not* directly compatible to fit onto a tractor with a Fast Hitch without using an adapter of some sort.

Aftermarket Cub Implements

There are dozens of other machinery manufacturers that built implements specifically for the Cub and Cub Lo-Boy. A very well-known example of this is Woods Manufacturing of Oregon, Illinois. They offered a line of mower decks for the Cub/Cub Lo-Boy tractors that looked similar to the ones that IH offered, yet they are uniquely different. Other manufacturers like Danco, Sunflower, Pennington, and others also offered implements to fit the Cub tractor line.

1950s CUB WITH FRONT GANG REEL MOWER
This unique mower setup has a gang reel mower placed in *front* of the Cub tractor. The front-mounted mowers of today must have been based off of this unit! The Cub has hydraulic lift to raise and lower the mower gangs.
Wisconsin Historical Society

FARMALL CUB POWERING GENERATOR
This Farmall Cub has a stationary generator of unknown manufacture attached to it. This is *not* the IH Electrall that was made by General Electric for IH that IH also offered. The tractor appears to be at some kind of IH exposition/display.
Wisconsin Historical Society

CHAPTER 6

Cub Paint and Decal Guide

The Farmall Cub is a simply painted tractor that is either a single solid color or two colors, depending on the age of the tractor. When production started, IH used #50 IH Red paint for the Cub. The tractor was nearly completely assembled before being painted, too! The starter, generator, gauges, and wiring were all installed before painting. The starter and generator ID tags, along with the gauge faces, were all masked off to not get paint on them.

IH also offered yellow-painted Cub and Cub Lo-Boys for those wanting or needing a brightly painted tractor for safety reasons. The industrial tractors were painted this way, with the yellow paint becoming the standard color in the 1960s, and red was the option. The yellow paint color that IH used was Federal Yellow. IH also offered the Cub/Cub Lo-Boy painted in Omaha Orange as a special order color for highway entities.

With the introduction of the two-color paint scheme in 1957, IH added #901 White to the tractors to be used on the tractor grille and (sometimes) wheels and wheel discs. These tractors have the stainless steel emblems, and the side of the hood is also white. IH did not have the time or manpower to paint the hood two colors, so they simply used a decal for the white background on the hood side (275001R and 275002R).

The new flat face (or flat grille) Cub and Cub Lo-Boy has a white-painted hood and grille, along with white-painted front wheel rims and center rear wheel discs. The rest of the tractor chassis is painted Federal Yellow. This yellow/white color combination would remain in use for the International Cub until its retirement in 1979. An all-red-painted version of the Cub would remain an optional color choice.

The "new" model 154 and 185 Cub Lo-Boy tractors also have this same yellow-and-white paint scheme.

The Model 184 Cub Lo-Boy saw the return to IH 2510 Red-painted body with IH 935 White-painted wheels and side engine trim panels.

IH used three different shades of white paint on the Cub/Cub Lo-Boy tractors.

Paint Code	Years Used
#901 White	1957–1967
#902 White	1967–1970
#935 White	1971–1979

The last printed IH owner's manual for the Cub tractor has (ironically) a red-painted tractor on the cover.

The US-built Cub and Cub Lo-Boy both used decals for their model identification until the mid 1950s when stamped stainless steel emblems appeared in 1955. The various caution/warning/instructional decals were always decals. They did not advance into stamped emblems. These decals did change into the self-stick (peel-and-stick) style in the 1960s.

The decals that IH used were of the water-transfer type, meaning that the decal was dipped in warm water and then slid (transferred) off the backing paper onto the machine. These worked flawlessly for decades until the use of self-stick (peel-and-stick) decals became common in the 1960s. The self-stick-style decals would be used until production of the Cub and Cub Lo-Boy tractors stopped in 1979.

In 1963, IH returned to using all decals only on the Cub and Cub Lo-Boy tractors instead of stamped-metal emblems. This would remain in effect until these tractors were retired in 1979.

ASSEMBLED CUB—READY TO SHIP

Still hanging from its overhead carrier, this new 1947 Cub has its hood decals freshly applied. Soon it will be test run and shipped to a waiting customer. *Wisconsin Historical Society*

TWO-TONE CUB

The 1958 model year Cub was given a two-color paint job to match the rest of the Farmall family. The white background on *all* of the tractors in this photo is a *decal*. Many times, restored tractors have this painted instead of using a decal like IH did. *Wisconsin Historical Society*

FARMALL CUB HOOD DECAL
The IH McCormick-Deering Farmall Cub decal on the hood of this tractor was fitted to only the very-early-built Cub tractors. Later tractors have the word "Deering" omitted. *Lee Klancher*

CUB LO-BOY DECAL
The 1968-made Cub Lo-Boy has a decal on the side of the hood. This same year, this tractor was replaced with the new model 154 Cub Lo-Boy. *Lee Klancher*

1950 CUB DEMO
The white paint of the 1950 Mid-Century Demo tractor offered very little color contrast to the decals. The lack of contrast almost makes the decals invisible. *Lee Klancher*

DASH DECALS ON TRACTOR
This view of the dash shows the yellow/black light switch decal, the "BE CAREFUL" caution decal and the harder-to-find Cub trademark decal. The Cub trademark decal was added to meet export qualities. *Lee Klancher*

CUB CHASSIS DECAL SET

This image displays the various caution, warning, and instructional decals used on the Cub and Cub Lo-Boy from 1955 to 1963. The patent decal was never offered in the tractor decal sets.
Kenneth Updike

INTERNATIONAL CUB DECAL
This photograph shows the International Cub hood decal used on the flat grille Cub tractors from serial number 222501 (a 1963 model) to serial number 248124 (a 1975 model). *Kenneth Updike*

INTERNATIONAL CUB LO-BOY DECAL
This hood decal was used on the flat grille 1963 to 1968 Cub Lo-Boy tractors. *Kenneth Updike*

McCORMICK FARMALL CUB DECAL
This hood decal was applied to the flat face Farmall Cub tractors built prior to serial number 224703. *Kenneth Updike*

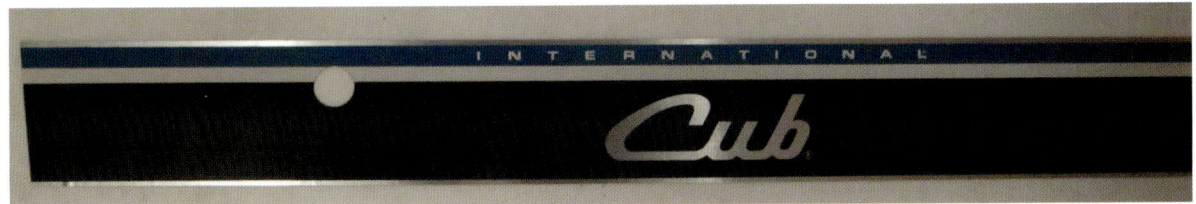

BLUE STRIPE CUB DECAL
Cub tractors built 1975 and later have a long blue stripe decal on their hood. This LH decal shows the circular cutout for the headlight mount. *Kenneth Updike*

CHAPTER 7

Repair and Maintenance

The real fun of owning vintage equipment is in the repair and maintenance, and the Farmall Cub is especially delightful to work on. If you take good care of your Cub, you'll find it to be a long-lasting and effective tool at your home, farm, or business. The Cub was designed to be easy to work on, and it's simple design has withstood the test of time. You don't need to be an experienced professional to keep your Cub running smoothly—you just need a bit of mechanical inclination, some ordinary shop tools, and a willingness to get your hands dirty. Let's get to work!

All repair and maintenance images by Jacob Hawkins unless otherwise credited.

7.1

How to Change Your Oil

An oil change on a Cub tractor is a simple task that is well within the grasp of anyone willing to turn a wrench and get a little dirt under their fingernails. Note that opinions vary a bit on the proper type of oil, but most use variable weight detergent oil with good results.

1 Position the drain pan under your tractor. Remove the drain plug. Let the oil drain into the pan.

2 Remove the bolt on top of the oil filter casing. The bolt is very long—you'll need to maneuver carefully to remove the bolt without taking off the hood of the tractor.

3 Lift off the cover and gasket, then pry out the filter. Some filters have an attached tab or wire to help with removal. If yours doesn't have one, use a screwdriver inserted into the hole to help lift the filter far enough out of the casing to get a grip on it. Wipe out the inside of the filter case. If the filter hasn't been changed in a while, the case interior will be very grimy.

4 Insert a new filter in the case. Put the end with the smaller hole down.

5

Replace the gasket on the casing cover.

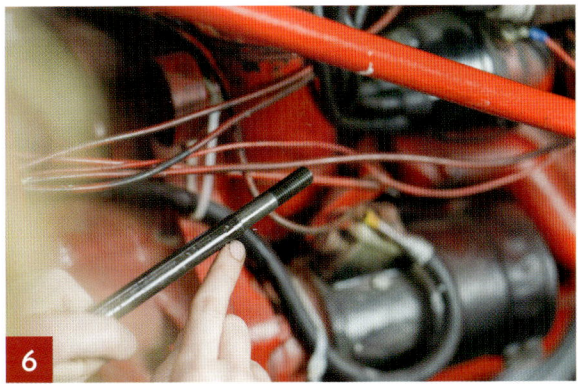

6

Insert the bolt through the cover and the filter. You'll need to wiggle the bolt around until you find the smaller hole at the bottom of the filter.

7

Carefully tighten the bolt. Because the bolt is hollow, it is easy to snap. Use just one hand on your wrench to prevent over-tightening.

8

Reinsert the drain plug and brass ring into the oil pan and tighten.

9

Since the air cleaner is an oil-bath system, you'll need to clean and change the oil in the air cleaner as well. Remove the cup that holds the oil by pulling the bale toward yourself.

10

Dump out the old oil and clean out the sediment using carburetor cleaner and a rag.

REPAIR AND MAINTENANCE

Refill the cup with oil to the line.

Reattach the cup to the air filter intake.

Pull the dipstick plug to open the oil tube.

Pour in a scant three quarts of oil. While an original manual would recommend low-ash 30W oil, we recommend the modern and readily available equivalent of 15-40 oil.

Measure with the dipstick to be sure the tractor is full, then replace the dipstick plug in the fill tube.

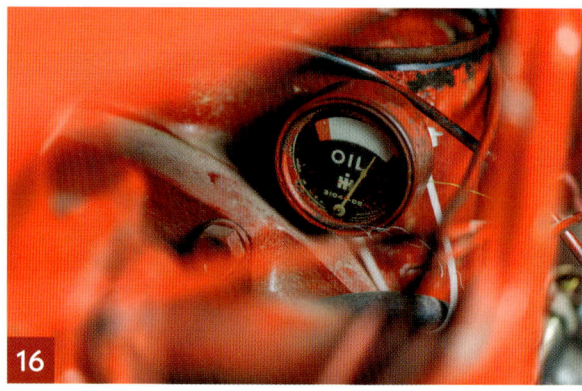

Start your tractor and check the oil pressure gauge. It should show thirty pounds of pressure within a few seconds of starting the tractor.

7.2

Lube and Grease

To flush the entire system, you'll need four quarts of 80/90 gear lube. However, a complete flush is rarely needed. Lube with a nozzle-top works best for this job.

Transmission Lube
Remove the plug on the top of the transmission casing. This is a tricky angle—use a crescent wrench and a screwdriver to turn the plug. Inspect the oil inside the transmission. It should be a deep brown or black color. If the fluid is milky, you'll need to drain and refill the entire system (drain plug is located at the bottom of the transmission housing). However, a complete flush is very rarely needed.

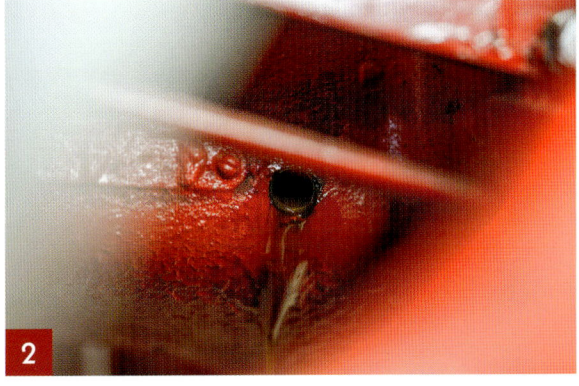

Remove the small level check plug on the side of the transmission housing to drain the fluid.

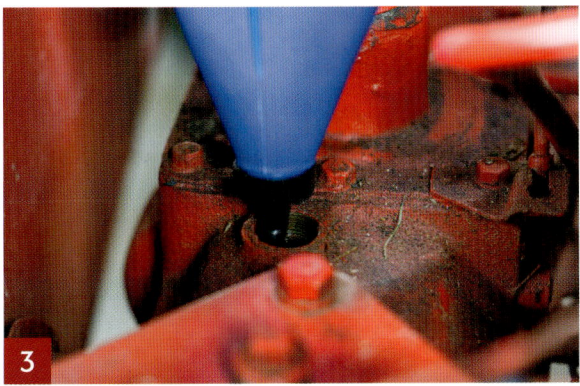

Slowly add oil through the hole on top of the transmission casing until oil begins to leak out of the level check hole.

Replace the plugs in the level check hole and on top of the transmission casing.

REPAIR AND MAINTENANCE

1

Final Drives
Remove the plug with the square head of the socket wrench to check the level.

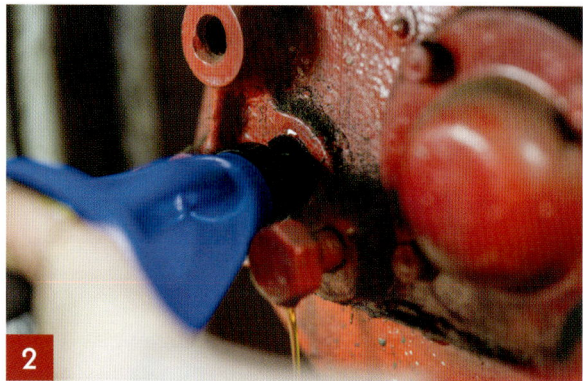

2

Insert the nozzle of the lube bottle into the hole, and add oil until oil starts to leak out of the hole.

3

Replace the plug and repeat the process on the other final drive.

4

Front End Grease
Use the grease gun to apply one pump of grease to each of these places:
1. Right tie rod end and spindle
2. Left tie rod end and spindle
3. Axle pivot
4. Center Steering Arm

7.3
Check the Air Pressure in Your Tires

Many Farmall Cub owners have added weight to their tractor by filling the rear tires with calcium chloride. The fluid can add weight to the tractor, helping with traction in tough conditions. To check the contents and pressure of your tractor's tires, follow these steps:

1 Position the tractor so that the rear tire's valve stem is easily accessible. Use a flathead screwdriver to gently but firmly press into the center of the valve stem to release a small amount of pressure.

2 If only air comes out, use a standard air pressure gauge to check the tire pressure, which should be between 10 and 15 psi. Refill if needed.

3 If liquid is released, use a water gauge to check the pressure inside the tire, which should be between 10 and 15 psi. If the pressure is low, refill by adding air.

4 Front tires are rarely filled with fluid. The sidewall of the tire should say what the air pressure should be (it varies significantly).

7.4

Adjust Your Brakes

When the tractor's brakes are functioning properly, the brake pedals should have about an inch of travel before they make contact. Your left brake and right brake should have an equal amount of travel so that they can be engaged simultaneously.

It's normal for brakes to slip out of alignment over time, but the repair is simple. Follow these steps to adjust your brakes:

Remove the cotter key and pin or hairpin clip (depending on what your tractor has). The cotter key can be wedged in tightly and may break as you remove it (be sure you have a new one handy before attempting the repair).

Remove the jam nut. Because this part is exposed, it often oxidizes in place. Don't be surprised if you need a torch to loosen it.

Turn the rod to shorten or lengthen, as needed. Three turns is roughly equal to an inch of play in the brake pedal.

Test by loosely reassembling the brakes and inserting the pin. Once you are satisfied with the adjustment, tighten the jam nut and secure the pin in place with the cotter key.

7.5

Adjust the Clutch Pedal

Just like with the brakes, a properly adjusted clutch pedal will have about an inch of free travel before it engages. Over time, tractors can fall out of adjustment. To reduce the amount of travel in your clutch, follow these steps:

These steps apply to late-style clutches which have an external adjustment bolt. Early Cubs have an internal adjustment mechanism which requires a different process to service.

1. Find the slotted lever where the clutch goes into the housing. Loosen (but do not remove) the cap screw that connects the rod to the lever.

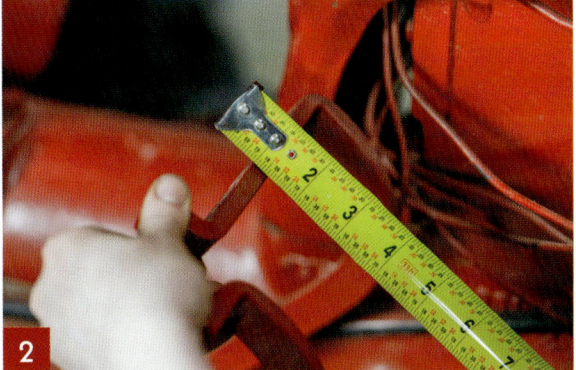

2. Slide the rod up or down as needed to achieve the correct amount of free travel in the clutch pedal, then tighten to secure.

7.6
Replace Your Front Seals

The steering column and front seal are often some of the least-maintained parts on a Farmall Cub. It's easy to forget, but a quick seal replacement and lube job will make a huge difference in the ease of steering the tractor. To get things fixed up, follow these steps:

Remove the nut from the end of the steering shaft. Sometimes this nut is stuck on—you might need to use a power tool to help remove it.

Use a gear puller to remove the wheel from the steering shaft.

Drive the half-moon key out of the steering shaft.

Unbolt the steering shaft bushing and lift it off of the steering shaft.

At the steering box, unbolt the steering box cover and pry it loose. Slide it up and off the driveshaft.

Use a screwdriver to tap or pry the old seal out of the steering box collar.

Gently tap a new double-lip seal in place. Use a socket to drive the new seal until it is flush.

Apply a fresh gasket to the steering box. Thread the seal collar back down the shaft and bolt it in place.

Reassemble the bushing, half-moon key, steering wheel, and nut. Consider adding a small amount of grease at the top of your shaft under the bushing.

Now, add oil to the steering box. Remove the top plug and inspect the lube inside the steering box. Like the transmission, it should be dark brown. If the fluid is milky, you'll need to flush the system completely (the drain plug is the lowest of the three plugs on the steering box). A complete flush is rarely needed.

Open the level check plug. It's the middle of the three plugs on the steering box.

Slowly pour 90W gear oil in through the top plug.

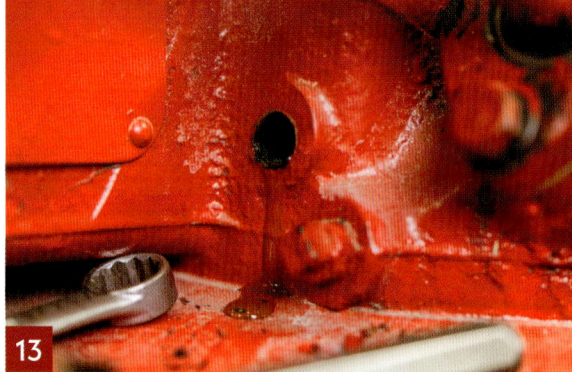

It is full when oil begins to flow out of the level check hole. Replace the plugs in the level check hole and the top of the steering box.

7.7
Rebuild Your Carburetor

A carburetor rebuild is the center of any good tune-up. Impurities will slip into the fuel system of even the most well-maintained tractor, gumming up the tiny channels in the carburetor and inhibiting the flow of fuel. An annual carburetor clean-out is necessary to remove these impurities, replace corroded parts, and restore the proper flow of fuel from the gas tank to the cylinders. If you know that you won't be using your tractor for six months or longer, consider adding stabilizer to the fuel before letting the tractor sit.

Please note that the following steps apply to an IH carburetor only.

Shut off the gas at the sediment bowl.

Loosen the bale and slide the bale to the side to release the glass bowl. Remove all remains of the old bowl gasket.

Replace the screen and gasket; clean the bowl and reinstall.

Remove the fuel line. Have a rag handy—there will be lots of dripping fuel.

5

Remove the bolts attaching the carburetor to the manifold and air intake hose.

6

Remove the choke and throttle linkages, using a pair of needle-nosed pliers to straighten and remove the cotter keys holding the linkages in place.

7

Once the carburetor is detached, carefully lift it off of the tractor and drain the fuel inside.

8

Loosen the four screws that hold the carburetor together.

9

Pull the top straight off to avoid breaking the idle tube.

10

Remove the float. You may need to use a small punch and pliers to remove the pin holding the float in place. Inspect the float for damage.

Remove the needle, seat, and accompanying gasket. A 7/16ths socket fits perfectly over the seat.

Remove the bowl gasket from the top of the carburetor.

Remove the emulsion tube and its gasket with a 5/16ths socket.

Remove the throttle shaft. Use a small screwdriver to remove the two screws holding the butterfly valve in place, then use pliers to remove the butterfly valve and the shaft underneath.

Remove the idle screw and the idle jet.

Remove the external gasket (where the carburetor connects to the manifold).

REPAIR AND MAINTENANCE 149

17 Use a small piece of wire to clean any sediment out of the main jet.

18 Clean the main jet and the main jet passage with a squirt of carburetor cleaner.

19 Use an air compressor fitted with a blow-off nozzle to blow out the main jet passage.

20 Clean all other holes and passages in the carburetor with carb cleaner and compressed air. This includes removing the old dust seal from the throttle shaft and any remaining gasket fragments.

21 Begin reassembly of the bottom of the carburetor by placing a gasket on the main jet.

22 Use a small amount of lubricant on the threads of the main jet, then screw it into place.

23 Inspect the choke door. It should open and close with a good seal.

24 Clean the discharge nozzle with a small piece of wire, then use a 5/16ths socket to screw it back into place with a new gasket.

25 Install a new throttle shaft and felt packing. Tap with the end of a screwdriver until flush.

26 Insert new throttle shaft. A small amount of lubricant helps.

27 Reattach the butterfly valve. The valve is directional, so check for markings indicating which side goes up. Twist to be sure it moves freely after installation.

28 Screw in the idle adjustment screw and the accompanying spring. Don't turn it in all the way—we'll adjust it once the tractor is running. For now, get it in about halfway.

29 Screw in the idle circuit screw and the accompanying spring. Turn it all the way in, then back it out one full turn.

30 Install the needle valve seat and its gasket, tightening firmly.

31 Drop the needle into the seat, with the pointed end downward into the seat.

32 Install a new bowl gasket, then reattach the float with a pin through the assembly.

33 Set the float. Use a ruler to measure 3/16th inch from the top of the float to the carburetor cover. Gently bend the float to achieve the correct set.

34 Screw the two halves of the carburetor together. Use a lock washer to secure each of the four screws.

Reattach the choke and throttle linkages. You'll likely need new cotter pins.

Gently reattach the fuel line. The threads are very easy to strip, so be careful not to over-tighten.

Attach the carburetor to the manifold (using a new gasket) and install the air intake hose.

Turn the gas on at the sediment bowl and start your tractor. While on a slow idle, turn the idle mixture screw until the tractor idles smoothly with no flutter.

Make an additional adjustment (while your tractor is still on a slow idle) to the fast idle screw, which is in the back and attached to the throttle shaft. Set this screw to your preference of speed of idle.

REPAIR AND MAINTENANCE

7.8

Service Your Battery and Cables

Make sure the battery and battery cables are in good shape. A clean connection will transmit the power your tractor needs for a smooth start.

1. Remove the ground battery cable.

2. Remove the plugs on the top of your battery (you could have three or more, depending on how many volts your electrical system is). Check the fluid level—it should be to the top of the cells.

3. If fluid is needed, use either distilled water or "electrolyte." I favor electrolyte because it doesn't freeze. Do not use regular tap water!

4. Clean the battery cable ends and posts. If the ends are bad, replace them.

7.9

Tune-Up Your Distributor or Magneto

The distributor or magneto is one of the most overlooked areas of regular maintenance. A tractor with a worn set of points will have a hard time starting and eventually not start at all. To keep your tractor in top condition, I recommend replacing the points each year. This schedule is especially necessary if your tractor is prone to corrosion (stored outside, or if you simply live in an area with a lot of rainfall). Tractors that are stored indoors and in arid climates may need less frequent maintenance. When servicing your tractor's distributor or magneto, you'll have a critical choice to make: Do you take the distributor or magneto off or leave it on the tractor? Removing the distributor or magneto makes it easy to see inside and replace the points, but it is surprisingly difficult to reinstall. A mistake in putting the distributor or magneto back on your tractor could throw your tractor out of time, resulting in a potential backfire or a failure to start altogether. If your tractor has a distributor, follow steps 1-16. If your tractor has a magneto, skip ahead to steps 17-25.

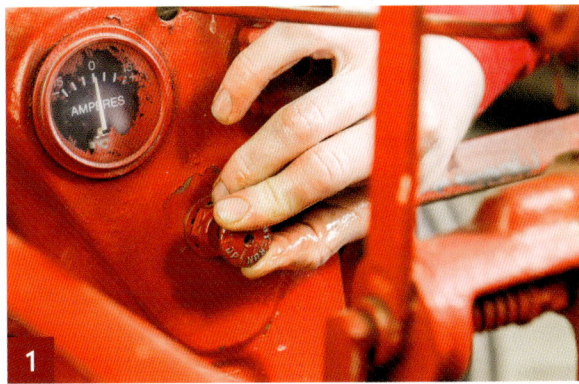

Turn off the power—use the switch on the dash.

Snap the two snaps off to remove the cap. (Note how the notch lines up between the cap and the distributor housing—you'll need to replace it the same way when you finish the repair.)

REPAIR AND MAINTENANCE

Inspect the inside cap. Check electrodes for signs of burning and wear.

For ease of repair, you may want to remove the distributor from the tractor. If you plan to do so, use the hand crank to turn the engine until the rotor points to the number 1 electrode. Otherwise, this is unnecessary.

Remove the rotor and dust cover. It will simply pull right off. The dust cover is important because it protects the points from dust, moisture, etc. Inspect for chips or cracks.

If you are removing the distributor from the tractor, this is where you do so. There are two bolts that hold the distributor in place. You'll also need to remove the coil wire.

Remove the old points and condenser from the inside of the distributor. A pair of pliers makes it easy to remove the points.

Put the condenser back in. Screw the condenser in tight, since the screw's connection with the back plate is what grounds the entire distributor.

9

Put the points in. Try to avoid touching the surface of the points. Securely tighten the nut that holds the condenser electrode and points together.

10

Rotate the distributor cam very slightly until it lifts the point to the highest position. If the distributor is out of the tractor, you can rotate this by hand. If the distributor is still in the tractor, use the hand crank to rotate. In either case, rotate it just enough to get the point into the high lobe position—be careful not to rotate more than necessary or your timing will be impacted.

11

Use a feeler gauge to determine the distance of the point gap. The point that screws into the base has a slot for easy adjustment. Once the gap is properly set, tighten the point into the base.

12

If you have removed the distributor from the tractor, put it back in now, but do not tighten it into place.

REPAIR AND MAINTENANCE

Replace the dust cover, felt washer, and the rotor. If you have removed the distributor, turn the distributor until the rotor points directly to the number 1 electrode on the distributor cap then tighten the bolts into place.

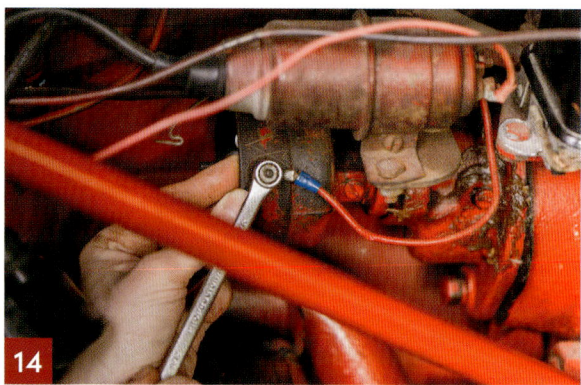

If you removed the coil wire, put it back on. Check to make sure all the attached wires are tight.

Put the cap back on, turn it till it clicks, and then secure it with the clip.

Start your tractor and listen to it run. If the tractor seems to labor when running, then slightly rotate your distributor to fine-tune the timing.

Some tractors were equipped with a magneto instead of a distributor. The photo on the left is of a magneto with an internal coil. Both magnetos and distributors have a cap, so inspect carefully to be sure your order the correct parts. If your tractor has a magneto, you can still replace the points and set the timing correctly.

To tune-up your magneto, remove the cap to expose the rotor. The rotor is held in place by a plate with two screws. Remove the screws and remove both the plate and the underlying gasket. Inspect the rotor carefully—if it is cracked or burnt, you'll need to replace it with a new one.

As you remove the rotor, take note of the small timing mark on the gear, which lines up with the mark on the rotor. You'll need to match these marks exactly during reassembly.

Next, remove the two screws that hold down the plate to the magneto housing. The gear will come off when you remove the plate.

With the plate removed, your points and condenser will be exposed. Replace the points and condenser following the same method used for replacing them in a distributor. Set the point gap between 12 to 15 thousands on high lobe.

Inspect the coil. A coil that is faulty will appear burnt or the contact will be damaged. To replace your coil, remove the screws from the side of the case to free up the bar that passes through the coil. Take note of the bar's position as you remove it (it is directional), and simply slide the old coil off and the new coil on to the same bar. Screw the bar with the new coil back in place. The coil wire is attached with the points.

REPAIR AND MAINTENANCE

23 After replacing your coil, points, and condenser, screw your silver plate back on and set your gear and rotor marks as shown.

24 Add some grease to the cavity, install a new gasket, and screw down the brown plate. Put the distributor magneto cap back on.

25 Add a few drops of oil into the magneto housing. Install it back onto the tractor just like the distributor above.

7.10
Service Your Charging System

(To watch a video tutorial of these troubleshooting steps, visit www.farmtractorrepair.com/cubcharging)

If your tractor needs a jump to get started, or if it suddenly quits and won't restart, you likely have a problem with your charging system. If your battery is old, it might need replacing—but if replacing the battery doesn't fix the problem, the trouble is likely somewhere in the charging system. Use these tests to determine exactly where the fault lies. To complete the troubleshooting, you'll need a momentary switch. This specialized tool allows you to momentarily ground out the generator field, bypassing the voltage regulator or cutout to find the problem in the charging system.

Test 1: Gauge Test
Start your tractor and look at the amp gauge. If the gauge shows a positive or climbing charge, the system is working as it should. If the gauge shows no change or a negative charge, something's not right. Proceed with the following tests to determine where the problem lies.

Test 2: Ground Cable Test
Loosen the ground cable on the tractor's battery then start your tractor. Once the tractor is running, remove the ground cable. A tractor in good repair ought to be able to keep running without being hooked up to your battery. If your tractor continues running well after the removal of the ground cable but failed the gauge test, you most likely have a faulty gauge. If the tractor quits once the ground cable is removed, then there's a problem with either the generator or the voltage regulator/cutout. This test applies only to tractors equipped with a distributor.

Test 3: Voltage Regulator/Cutout Bypass Test

Voltage regulators and voltage cutouts serve the same basic purpose: to cut off the feed of electricity to the battery once the battery is fully charged. Early Cub tractors were originally equipped with a simple voltage cutout (one coil under the cover; serial number 221292 and earlier). Later models were equipped with a voltage regulator (two coils under the cover). However, I've seen plenty of Cubs that have been retrofitted with a different system than they left the factory with, so if this test fails, you'll simply need to be sure you get the correct new part.

Whether your tractor is equipped with a voltage regulator or a voltage cutout, the instructions are the same.

If the cutout or regulator is faulty, it will prevent an operational generator from fully charging the battery.

To begin the test, clip one end of the momentary switch to the field terminal (marked with the letter "F").

Next, start the tractor. Touch the other end of the momentary switch to ground and press the switch.

Watch for activity on the tractor's amp gauge. If the amp gauge shows activity, your generator is working properly and the fault lies in your cutout or regulator (it's an easy replacement). If there is no activity on the gauge, the generator is faulty. I recommend taking the generator to a specialty shop for repair.

Test 4: Armature-to-Power/Field-to-Ground Test

This test can be performed to double-check test number 3. Since a generator repair or replacement can be quite costly, I recommend taking the extra time to be sure the generator is the problem. This test can only be performed on a 6-volt generator.

1. Disconnect the belt from the generator. Use a jumper wire to connect the field terminal on the generator (marked with an "F") to a ground connection near the generator.

2. Take the wire off of the battery connection on the voltage regulator/cutout and touch it to the "A" terminal on the generator. An operational generator will spin smoothly—this points to a problem with your voltage regulator/cutout. If the generator does not begin to spin, your generator is faulty and needs to be rebuilt or repaired. Troubleshooting note: It is entirely possible for an aging electrical system to have more than one faulty part. You may find that after repairing your generator the system still does not charge. If that's the case, repeat test number 3 to see if you also have a problem with your voltage regulator or cutout. Also check all of the connections (they often get rusty) and the high-low dash switch.

7.11

Replace Your Spark Plugs

Worn spark plugs are the most common cause for engine misfires—but even before they erode to the point where they cause misfires, they can make the engine run rough and with poor fuel economy. Spark plugs are an easy and inexpensive way to improve engine performance. A Farmall Cub will need four (one for each cylinder) D-18Y Champion or equivalent plugs. To replace the plugs:

Remove the wires leading to the spark plugs. Thoroughly inspect for damage—if the end is broken or corroded, it will need to be replaced.

Unscrew the plugs from the engine block with either a deep well socket or wrench.

Use a feeler gauge to check the gap on the plugs. The gap should be 0.025 inches. If the gap is larger or smaller, make an adjustment before installing.

Reinstall the spark plugs and tighten snugly. Reattach the spark plug wires.

7.12

Replace the Starter Switch and Button

If you pull the starter ring and the tractor acts like it's not kicking in or you have to pull it extra hard, it's time to replace the starter button. It's normal for the copper contact inside the starter to corrode over time. The repair is easy and inexpensive—here's how:

Remove the power cable from the battery.

Remove the other end of the battery ground cable and the ignition wire from the starter button. Then remove the starter actuator and the two screws that hold the button onto the starter.

Install a new starter button, making sure to use the same screws (they are very short—long screws would go through the case and hit the armature or fields to damage the starter). Reconnect the actuator, ignition wire, and battery cable to the starter.

Finally, reconnect the battery cable.

7.13

Test Your Coil

This test is for tractors equipped with a distributor. A sure sign of a faulty coil is a tractor that runs for around twenty minutes and then suddenly quits. If your tractor does this and won't restart until it has cooled down, it's very likely that your coil is the problem. To confirm, perform this test:

1 Touch both the positive and the negative electrodes on the back of the coil with both ends of the ohmmeter.

2 If the 6-volt coil is good, the readout should be between 1.3 and 1.5 ohms. If the readout is inconsistent or lower/higher than it ought to be, your coil is weak and needs replacing.

7.14

Test Your Engine Compression

If your tractor seems to run weakly or smokes, the problem likely lies with worn valves. You can easily diagnose trouble with the valves by performing a compression test.

You'll need a special gauge to perform the test. Compression test gauges are available in two styles: ones that screw in to the spark plug hole (with adapters for various sized spark plugs) and ones that use a rubber nozzle that you hold tightly into the hole to form a seal. The results will be far more reliable with the screw-in version.

1 To perform a compression test, simply remove a spark plug and screw in the gauge. Press the starter button and let your tractor roll over a couple of times before taking the reading.

2 A healthy engine will show a reading over 100 psi—ideally, between 110 and 120. If your gauge reads below 100, your tractor is ready for a valve adjustment or a complete overhaul. Repeat this process for every cylinder, as valves can wear out at uneven rates.

7.15

Adjust Your Engine Valves

(Watch a video tutorial of this repair at www.farmtractorrepair.com/cubvalves)

Adjusting the valves in your tractor's engine optimizes the seal on each cylinder, giving your tractor the compression it needs for maximum performance. It's normal for valves to wear down and to loosen over time. While tractors will need periodic engine rebuilds to replace truly worn-out valves, an annual valve adjustment can go a long way toward improving the performance of your tractor between rebuilds.

Remove the carburetor and set it aside.

Remove the valve tappet cover. You may need to use a screwdriver to pry it off. With careful maneuvering, the cover can be removed without needing to remove the hydraulic lines and manifold.

Remove spark plug number 1 (closest to the radiator).

Put your finger on top of the hole where the number 1 spark plug goes. Roll the tractor over using a hand crank until you feel a small puff of air.

5 Crank slowly until the top dead center notch on the fan-drive pulley is aligned with the pointer.

6 Adjust valve numbers 1, 2, 3, and 5 (number 1 is closest to the radiator) by inserting a 0.015-inch feeler gauge between the valve tappet and the valve stem. Turn the adjusting screw (this requires two wrenches) to make the gap smaller or larger as needed. Hint: Turn clockwise for a bigger gap, or turn counter-clockwise for a smaller gap.

7 Repeat steps 4 and 5 with the number 4 spark plug. Once the fourth cylinder is in top dead center (one engine revolution), adjust valves 4, 6, 7, and 8.

8 Replace the valve tappet cover, using a fresh gasket. Reinsert the spark plugs and reattach the carburetor.

APPENDICES

Cub Serial Number Listings

Farmall Cub Tractor
Prefix Letters FCUB

501 to 11347	Built in 1947
11348 to 57830	Built in 1948
57831 to 99535	Built in 1949
99536 to 121453	Built in 1950
121454 to 144454	Built in 1951
144455 to 162283	Built in 1952
162284 to 179411	Built in 1953
179412 to 186440	Built in 1954
186441 to 193657	Built in 1955
193658 to 198230	Built in 1956
198231 to 204388	Built in 1957
204389 to 211440	Built in 1958
211441 to 214973	Built in 1959
214974 to 217381	Built in 1960
217382 to 220037	Built in 1961
220038 to 221382	Built in 1962
221383 to 223452	Built in 1963
223453 to 224703	Built in 1964

International Cub Tractor

224704 to 225109	Built in 1964
225110 to 227208	Built in 1965
227209 to 229224	Built in 1966
229225 to 231004	Built in 1967
231005 to 232980	Built in 1968
232981 to 234867	Built in 1969
234868 to 236826	Built in 1970
236827 to 238505	Built in 1971
238506 to 240580	Built in 1972
240581 to 242745	Built in 1973
242746 to 245650	Built in 1974
245651 to 248124	Built in 1975

International Cub Lo-Boy Tractor

501 to 2554	Built in 1955
2555 to 3928	Built in 1956
3929 to 6581	Built in 1957
6582 to 10566	Built in 1958
10567 to 12370	Built in 1959
12371 to 13903	Built in 1960
13904 to 15505	Built in 1961
15506 to 16439	Built in 1962
16440 to 17927	Built in 1963
17928 to 19405	Built in 1964
19406 to 21175	Built in 1965
21176 to 23114	Built in 1966
23115 to 24480	Built in 1967
24481 to 26007	Built in 1968

International Cub 154 Lo-Boy Tractors

7505 to 8272	Built in 1968
8273 to 15501	Built in 1969
15502 to 20331	Built in 1970
20332 to 23342	Built in 1971
23343 to 27537	Built in 1972
27538 to 31765	Built in 1973
31766 to 36676	Built in 1974

International Cub 185 Lo-Boy Tractors

37001 to 37315	Built in 1974
37316 to 41240	Built in 1975
41241 to 43347	Built in 1976

International Cub Tractor (with Increased HP)

248125 to 248617	Built in 1975
248618 to 250831	Built in 1976
250832 to 252108	Built in 1977
252109 to 253135	Built in 1978
253136 to 253685	Built in 1979

International Cub 184 Lo-Boy Tractors

43802 to 46162	Built in 1977
46163 to 48029	Built in 1978
48030 to 52339	Built in 1979

Model	Engine	Engine HP	Transmission Type	List Price
154	C-60	15	Gear	$2,405
185	C-60	18	Gear	$4,295
184	C-60	19	Gear	$5,650

ACKNOWLEDGMENTS

This book would not have been possible without the contributions of my family. My dad, Dan Gingell, gave me the great gift of mechanical know-how and continues to be my best teacher. My mother, Jennifer Gingell, supported this endeavor and every one of my entrepreneurial ventures from the very start. My sister, Elizabeth McAdams, helped put my thoughts to paper and made this a reality.

Roger Fortier and Ian Campbell, both gifted mechanics and technical experts, painstakingly read over each tutorial and checked them for accuracy.

Steve and Joanne Mason, longtime friends and owners of the most beautiful farm in the neighborhood, allowed us to tromp all over their land to take pictures.

Jacob Hawkins, photographer extraordinaire, is gifted with endless patience and great artistic skill. His vision has brought this book to life.

Isabella Hollo and Ena McAdams, my favorite little mechanics, inspire me to teach the next generation.

—Rachel Gingell

To give a sincere thank you to everyone who helped shape this book would be an understatement. To thank everyone in any kind of order would be inappropriate, but a few individuals do deserve mention in this book.

Thank you once again to the friendly staff of the Wisconsin Historical Society and especially employees Guy Fay and Lee Grady. These two individuals have been charged with the gargantuan job of safekeeping and inventory of the IH/McCormick collection at the State Archives. Finding boxes of sales literature, photos, and production data for me that quickly went to the photocopier (for later study) made this book research go much easier. I thank you both.

I need to mention thanks to my grandparents Harvey and Alice Kilian. My exposure to IH in my early childhood was often spent driving their Model H or Model M Farmalls in the machine shed along with my cousin Doug. They have all passed on now, but I know this preserves my memories of them. I blame them for getting me hooked on IH.

A special note of thanks to my parents (Ed and Cathy), who never minded my collection of IH toy tractors and literature, being stored at their house (even if it took up a lot of their basement).

The staff at Johnson Tractor (the Case IH dealer I work for) have been extra flexible with me. Their cooperation in my research and fact-finding was exceptional. The experiences I have working there every day are incredible.

My thanks also extend to Jeff Powell for his expert research. Thank you to Jim Becker for the outstanding clues on Cub production and features. A thank you also to Mike Schmudlach, who imported several French Cubs and shared them with me and the rest of the United States.

Thanks to the great folks at www.farmallcub.com. They keep and share the history of the Cub tractor online. To produce a book with ideas and input from them is a real joy. I'm sure that readers will always want *more*, but that is what sequels are for.

I must say thanks to my friend and publisher Lee Klancher at Octane Press (who published this book). I am glad we finally got this book back on the rails. Thanks for all of your help in this.

Finally, thanks to my co-author on this project Rachel Gingell. She was tasked with covering the how-to sections and has done a wonderful job.

—Kenneth Updike

INDEX

1947 Harvester Farm exhibit, 11
1947 Cub, 16, 18, 29, 34, 130
1950 White Demo C, 34, 36
1950 White Demo Cub, 34, 35-38, 122, 131
1950 White Demo Cub, determining fakes, 36
1950 White Demo Super A, 34, 36
1955 Cub, 29, 47
1955 Cub development, 47
1956 Cub, 47
1957 Cub Lo-Boy, 120
1958 Cub, 130
1959 Cub, 52, 55
1959 Cub development, 52
1959 Cub Lo-Boy, 52
1960 Cub, 55
1960 Cub Lo-Boy, 52
1961 Cub, 55
1962 Cub, 55
1963 Cub, 53, 55, 120, 133
1963 Cub Lo-Boy, 133
1964 Cub, 56, 57
1964 Cub development, 56
1964 International Cub, 57
1968 Cub Lo-Boy, 57, 131, 133
1968 Model 154 Cub Lo-Boy, 131
1975 Cub, 72, 133
1978 International Cub, 72
1978 Model 184 Cub Lo-Boy, 72
1978 Model 284 Cub Lo-Boy, 72
1979 International Cub, 72
1979 International Cub cost, 72
AC-Delco, 45, 80
Allied Equipment, 40
Allis-Chalmers, 42
Allis-Chalmers Model 720, 82
Allis-Chalmers Model G, 19
Allis-Chalmers Onan engine, 82
Bolens Kohler engine, 82
Bolens Model 1250, 68
Bolens Model 1455, 68
Bolens Model HT 23, 82
Boy Scouts of America, 36
Calcium chloride, for wheel weight, 141
Canton Works, 23, 29, 30, 111
"Carousel Bear," 11
Chicago Museum of Science and Industry, 11

CNH Industrial, 92
Croix factory (France), 86
Cub Cadet, 7, 21, 26, 27, 51, 63, 64, 68, 80, 103
Cub Cadet 82 series, 83
Cub Cadet 122, 66
Cub Cadet 123, 66
Cub Cadet 124, 66
Cub Cadet 125, 66
Cub Cadet Corporation, 26
Cub cost, 11, 12, 37, 38, 57, 122
Cub demonstrations, 16, 34, 36, 39, 58, 127
Cub design, 11, 12, 15-19, 24
Cub, end of Farmall version production, 57
Cub, first prototype, 11
Cub parts books, 42, 97
Cub promotions, 11, 33, 36, 37, 58
Cub, total number built, 27, 72
Cub Industrial, 129
Cub Industrial development, 58
Cub Industrial, date dropped from IH Industrial line, 62
Cub Industrial Lo-Boy, 58, 59
Cub Lo-Boy, 7, 12, 21, 24, 26, 39, 40-47, 49-52, 57-60, 62-67, 69, 92, 95-98, 101, 103, 105, 111, 112, 120-122, 127, 128, 130, 132
Cub Lo-Boy cost, 41, 46
Cub Lo-Boy development, 39, 40
Cub sales literature, 65, 67, 70, 78
Cub serial number 501, 12, 29, 30
Cub serial number 502, 29, 34
Cub serial number 503, 29, 34
Cub serial number 504, 29, 34
Cub serial number 505, 29, 34
Cub serial number 506, 29, 34
Culti-Vision, 31, 33, 111
Culti-Vision A, 11
Culti-Vision AV, 11
Curtis Aircraft Company, 20
Danco, 58, 127
Date of first Cub built, 30
Date of last Cub built, 27, 72
David Bradley Company, 124
Delco Remy Company, 22, 37, 64, 96
Dresser Industries, 27
Ducellier, 84

Farmall 140, 27, 120
Farmall 300, 47
Farmall 400, 47
Farmall A, 11, 21, 78, 98
Farmall A1, 78
Farmall AV, 78
Farmall AV-1, 78
Farmall B, 11, 21, 78, 98
Farmall BN, 78
Farmall C, 78, 104
Farmall F-12, 11
Farmall F-14, 11
Farmall H, 12
Farmall M, 12
Farmall Super A, 12, 27, 78, 104
Farmall TR3, 27
Farmall TR4, 27
Farmall X, 11
Farm Tractor Engineering Department, 39, 40
Ferguson system, 41, 42
Final production Cub (serial number 253685), 27
Firestone tires, 103
First production Cub (serial number 501), 12, 21, 29, 30
First production Cub Lo-Boy (serial number 501), 42
Ford 2N, 42
Ford 9N, 42
Ford 8N, 42
French-built Cub (FFCUB), 84, 86, 87, 95
French-built Cub Narrow, 86
French-built Super Cub (FSCUB), 84, 86
General Electric, 127
General Motors, 45
General Sales Department, 41
Godfrey, Arthur, 33
Goodrich tires, 103
Goodyear tires, 103
Gruhn Hybrid Corn, 29
GUIDE Light Company, 45
Hamilton factory, 111
Harris, J.E., 29, 30
Harvester Experimental Farm, 16, 18
Hinsdale demonstration program, 16, 18, 39

Hinsdale Engineering Center, 75
Holley, 95
Hundred Series tractors, 47, 72
Hydro Cub Lo-Boy concept, 68, 83
IH 50 series, 83
IH 86 series, 76, 83
IH 200 series, 72
IH Farm Equipment Committee Report number 249, 63
IH Industrial Equipment, 58
IH Works (Mexico), 29
Implement pricing, 1950, 122
Implement Works, 40
Indianapolis Works, 23
International 184, 82
International 460, 52
International 560, 52
International 600, 41
International 650, 41
International 686, 76
International 982, 82
International 4100, 67
International Cub, total number built, 72
International Harvester (IH), 7, 9, 11, 12, 14-16, 18-24, 26, 27, 29-31, 33-43, 45-47, 49, 51, 52, 54, 56-58, 62-73, 75-77, 79, 80, 82-84, 86, 87, 89-92, 94-98, 100-105, 111, 120-124, 127, 129, 130, 147
J.I. Case Model 442, 68
J.I. Case Model 444, 68
John Deere Kohler engine, 82
John Deere Model 140, 68
John Deere Model 400, 82
Knudsen Implement Co., 29
Letter Series, 9
Lo-Ash oil, 92
Louisville Airport, 27
Louisville Works, 16, 20, 21, 23, 26, 27, 29, 30, 34, 36, 65, 84, 86, 89
Louisville Works, closure, 26, 27
Louisville Works, creation, 20, 21, 23
Marvel-Schebler, 95
McCormick, Fowler, 16
McCormick Works, 23
Memphis factory, 111
Mid-Century demo program, 24, 34-36, 131
Miller True Value Hardware, 27
Milwaukee Works, 23
Minnesota Department of Transportation (DOT), 59
Model 154 Cub Lo-Boy, 57, 63-69, 71, 72, 79, 103, 120, 121, 129
Model 154 Cub Lo-Boy, total number built, 68
Model 154 Cub Lo-Boy, body styles, 68
Model 154 Cub Lo-Boy, cost, 68

Model 154 Cub Lo-Boy High-Clearance, 68
Model 154 Cub Lo-Boy prototype, 63
Model 184 Cub Lo-Boy, 26, 64, 68, 76-82, 97, 103, 121, 129
Model 184 Cub Lo-Boys, total number built, 83
Model 185 Cub Lo-Boy, 64, 68-71, 76, 78, 80, 83, 103, 121, 129
Model 185 Cub Lo-Boys, total number built, 69, 71
Model 185 Cub Lo-Boy, cost, 71
Montaire factory (France), 86
MTD Products Inc., 26
Nebraska Tractor Testing facility, 68
New International Cub Tractor series, 72
Paris Rhone, 84
Pennington, 127
Piper & Paine Manufacturing, 60, 61, 62
Purolator, 92
Rock Falls Works, 23
Sears, 124
Simplicity Model 9020, 82
Simplicity Onan engine, 82
Sunflower, 127
Tractor Committee Report Number 370, 39
Tractor Committee Report Number 141, 57
Tractor Works, 23
United Parcel Service (UPS), 27
War Assets Administration, 20
West Pullman Works, 23
Wheel Horse Model GT 14, 68
Woods Manufacturing, 127
World War II, 7, 12, 15, 16, 19, 20, 94
Zenith, 69, 70, 76, 95

Parts and Systems
Air cleaner, 35, 64, 137
Alternator, 80, 97
Battery, 44, 45, 64, 96, 97, 108
Battery, servicing, 154
Belt pulley, 37
Bendix starter, 64, 96
Brakes, 37, 56, 71, 80, 81, 83
Brakes, adjusting, 142
Brakes, inspecting, 108, 109
C-60 engine, 12, 20-22, 35, 37, 41, 43, 48, 63, 64, 66, 69-72, 76, 83, 86, 89-92, 95, 98, 100
C-60 engine, inspecting, 108, 109
Cables, servicing, 154
Carburetor, 35, 54, 69, 70, 76, 90, 94, 95
Carburetor, rebuilding, 147-153
Charging system, servicing, 161-163
Clutch, 15, 63-65, 80, 83, 100

Clutch, adjusting pedal, 143
Clutch, Aubern brand, 100
Clutch, inspection, 108
Clutch, Rockford brand, 100
Cooling system, 63, 64, 76, 98, 99
Counterbalance weights, 62
Cowling, 41
Dash panel, 49, 55, 60, 66, 75, 131
Decals, 18, 29, 34, 42, 43, 45, 47, 57, 58, 62, 68, 72, 76, 81, 93, 129-133
Differential, 41
Distributor, 64, 91, 96
Distributor, testing coil, 166
Distributor, tuning up, 155, 156-158
Draw bar, 37, 67, 113, 114
Electrical system, 45, 46, 55, 64, 75, 80, 86, 91, 96, 97
Electrical system, inspecting, 108
Electric start, 22, 37, 64, 78, 80, 89, 96
Emblems, 42, 43, 45-47, 50, 55, 84, 86, 129
Engine compression, testing, 167
Engine valves, adjusting, 168, 169
Exhaust, 40, 41, 54, 57, 64, 70, 78, 84, 87, 90
Fast Hitch, 40, 41, 46, 47, 72, 84, 104, 111, 112, 122-126
Fenders, 37, 66
Final drives, 47, 63, 100, 101
Final drives, adding lubricant, 140
Frame, 63, 66, 68, 78, 80, 83
Front axle, 37, 39, 47, 64, 77, 102
Front end, adding grease, 140
Fuel system, 94
Fuel system, filter, 94
Gas cap, 48, 81
Gas tank, 48, 82, 104
Gearbox, 80, 83, 100
Gearshift lever, 39, 61, 80, 83, 101
Generator, 22, 46, 64, 78, 80, 97
Grille, 18, 19, 42, 43, 47, 50, 52, 56, 57, 68, 69, 71, 72, 76, 79, 84, 86, 98, 129, 133
Hand crank, 97, 108, 109
Hood, 43, 45, 48, 55, 62, 68, 69, 71, 72, 76, 129
Hood, inspecting, 108
Horn, 40
Hour meter, 63
Hydraulic system, 61, 65-67, 70-72, 86, 89, 104, 105, 120
Hydraulic system, inspecting, 108
Hy-Tran Ultraction fluid, 104
Ignition, 37, 64, 80, 96
Ignition key, 80
License plate, 87
Lighting, 22, 32, 37, 41, 45, 53, 64, 69, 72, 96, 97
Magneto, 37, 96, 97
Magneto, tuning up, 155, 158-160

Muffler, 37
Oil system, 92, 93, 96
Oil system, changing oil, 136-138
Oil system, determining type of oil used previously, 92
Oil system, filter, 67, 92, 93, 96, 136
Oil system, inspecting, 108
Paint, 22, 23, 27, 34, 36, 42, 43, 45, 47, 52, 55-59, 68, 69, 71, 72, 76, 87, 95, 103, 129, 130
Paint, inspecting, 108, 109, 120
Power take-off, 37, 45, 59, 63, 65, 66, 71-73, 80, 82, 118
Radiator, 18, 35, 38, 63, 64, 76, 78, 83, 98, 99
Rear axle, 39, 66, 102, 120
Rear hitch, 66, 67, 70, 82, 83, 117, 122, 126
Seals, replacing front, 144
Seat, 37, 39, 44, 54, 60, 61, 66, 68, 72, 101
Serial number plate, 51, 71, 85
Service history, inspecting, 109
Sheet metal, inspecting, 108
Spark arrestor, 37
Spark plugs, replacing, 164
Starter switch and button, replacing, 165
Steering, 38, 60, 64, 71, 83, 120
Steering box, adding lubricant, 146
Steering box, seal replacement, 145
Steering shaft, seal replacement, 144, 145
Suspension, 84, 113
Tires, 37, 66, 73, 102, 103
Tires, checking air pressure, 141
Tires, checking contents, 141
Tires, inspecting, 108, 109
Touch Control, 29, 30, 36, 37, 67, 72, 74, 104, 105, 113, 118
Touch Control, inspecting, 108
Touch Control block, 104, 105
Touch Control fluid, 104, 105
Transmission, 37, 41, 64, 65, 66, 68, 71, 80, 83, 100, 101
Transmission, changing lubricant, 139
Transmission, inspecting, 108
Transmission, performing overhaul, 101
Universal Mounting Frame, 118
Wheels, 37, 42, 56, 66, 71, 72, 76, 79, 102, 103, 129
Wheels, inspecting, 108
Wheel weights, 37, 51, 102, 103

Implements
Auger drill, 25
Bean harvester, one-row, 112
Bean harvester, two-row, 38
Buzz saw, rear-mounted, 14
Cable trencher, 59
Cultivator, 6, 13, 17, 30, 31, 33, 38, 53, 54, 84, 111, 122
Cultivator, one-row, 112, 119
Cultivator, four-row, 112, 119
Cultivator, spring-tooth, 112, 122
Cultivator, two-row, 112, 118
Digger, 47
Digger, Fast Hitch, 124
Disc harrow, 32, 38, 41, 111
Disc harrow, bush and bog, Fast Hitch, 126
Disc harrow, Fast Hitch, 122, 125, 126
Disc harrow, Lo-Boy Fast Hitch, 112
Disc harrow, single, 112, 115
Disc harrow, tandem, 112, 114, 122
Disc harrow, tandem, Lo-Boy numbered series, 121
Disc harrow, tandem, Lo-Boy numbered series, 121
Disc plow, 38, 111, 113, 114
Disc plow, Fast Hitch, 112
Disc plow, one furrow, 112-114, 122
Disc plow, one furrow with Timken roller or Chilled bearing, 112
Farm trailer, 38, 77, 112
Farm trailer, Lo-Boy numbered series, 121
Fertilizer applicator, 6, 32, 33
Forklift, 60-62
Generator, IH Electrall, 127
Grader blade, 13, 17, 24, 38, 57, 74, 76, 101, 104, 105, 112, 122
Grader blade, Lo-Boy, 112
Grader blade, Lo-Boy numbered series, 121
Harrow, peg-tooth, Fast Hitch, 126
Harrow, peg-tooth, Lo-Boy Fast Hitch, 112, 115
Harrow, four-bar, spike-tooth, 121
Harrow, spring-tooth, Fast Hitch, 126
Harrow, spring-tooth, Lo-Boy Fast Hitch, 112
Harrow plow, two-disc, 112
Hay loader, 117
Loader, Model 1000 front-end, 44, 46, 53, 55, 112, 120
Loader, front-end, Lo-Boy numbered series, 121
Manure spreader, 42
Middlebuster, 112
Middlebuster, Fast Hitch, 125
Moldboard plow, 18, 25, 38, 41, 52, 86, 111, 112, 122
Moldboard plow, Fast Hitch, 125
Moldboard plow, Lo-Boy Fast Hitch, 112
Moldboard plow, Lo-Boy numbered series, 121
Moldboard plow, one-furrow, two-way, 112, 113
Moldboard plow, one-furrow, one-way, 112, 113
Mower deck, 26, 65, 66, 72, 73, 86
Mower deck, Lo-Boy numbered series, 121
Mower, finish, Lo-Boy numbered series, 121
Mower, flail, 49, 60
Mower, flail, Fast Hitch, 124
Mower, gang reel, Lo-Boy numbered series, 127
Mower, rotary, 30, 63
Mower, rotary, Lo-Boy numbered series, 121
Mower, sickle-bar, 34, 38, 52, 59, 63, 79, 112, 118, 122
Mower, sickle-bar, Lo-Boy numbered series, 121
Planter, 32, 111, 119
Planter, Fast Hitch, 122, 123
Planter, backland, 112
Planter, one-row, 112
Planter, one-row, Fast Hitch, 123
Planter, one-row backland cotton, 112
Planter, one-row cotton, 112
Planter, runner, 112
Planter, two-row, 18
Planter, two-row drill-type, 112
Platform carrier, 41
Platform carrier, Fast Hitch, 122, 123
Rake, side-delivery, 117
Rear scoop, Lo-Boy Fast Hitch, 124
Rotary hoe, Fast Hitch, 125
Rotary hoe, Lo-Boy Fast Hitch, 112
Snow blower, 58, 80
Snow thrower, Lo-Boy numbered series, 121
Snow plow, 81
Soil pulverizer, 116
Tiller, 80
Tiller, Lo-Boy numbered series, 121
Tool bar, rear-mounted, 38, 112, 122
Vegetable planter, 19, 118
Vegetable planter, four-row, 112, 122
Vegetable planter, one-row, 112
Weeder-mulcher, 116

Octane Press, Edition 1.2, May 2025
Edition 1.1, December 2021
Edition 1.0, July 2019
© 2019 by Kenneth Updike and Rachel Gingell

All rights reserved. With the exception of quoting brief passages for the purposes of review, no part of this publication may be reproduced without prior written permission from the publisher.

ISBN: 978-1-937747-25-1

Cover and Interior Design by Tom Heffron
Project edited by Aki Neumann
Copyedited by Dana Henricks
Proofread by Peter Schletty

On the Cover: Author Rachel Gingell at work in her shop, on her Farmall Cub. *Jacob Hawkins*

On the Frontispiece: The distinctive grill of the iconic Cub. *Jacob Hawkins*

On the Title page: This Cub was photographed not far from the Gingell family farm. *Jacob Hawkins*

On the Contents page: Farmall Cub. *Jacob Hawkins*

On the Foreword page: This Cub is cultivating and applying fertilizer. *Wisconsin Historical Society*

On the Introduction page: Author Rachel Gingell in her shop in Michigan. *Jacob Hawkins*

octanepress.com

Printed in China